Microscale Techniques For The Organic Laboratory

Second Edition

Microscale TECHNIQUES FOR THE ORGANIC LABORATORY

Second Edition

Dana W. Mayo
Bowdoin College

Ronald M. Pike
Merrimack College

Peter K. Trumper
University of Maine School of Law

JOHN WILEY & SONS, INC.
New York   Chichester
Brisbane   Toronto   Singapore

All experiments contained herein have been performed several times by students in college laboratories under supervision of the authors. If performed with the materials and equipment specified in this text, in accordance with the methods developed in this text, the authors believe the experiments to be a safe and valuable educational experience. However, all duplication or performance of these experiments is conducted at one's own risk. The authors do not warrant or guarantee the safety of individuals performing these experiments. The authors hereby disclaim any liability for any loss or damage claimed to have resulted from or related in any way to the experiments, regardless of the form of action.

"Permission for the publication herein of Sadtler Standard Spectra ® has been granted, and all rights reserved, by Sadtler Research Laboratories, Division of Bio-Rad Laboratories, Inc."

ACQUISITIONS EDITOR   Jennifer Yee

MARKETING MANAGER   Charity Robey

SENIOR PRODUCTION EDITOR   Patricia McFadden

SENIOR DESIGNER   Dawn Stanley

PRODUCTION MANAGEMENT SERVICES   Ingrao Associates

This book is printed on acid-free paper. ∞

Library of Congress Cataloging-in-Publication Data

Mayo, Dana W.
    Microscale technique for the organic laboratory/Dana W. Mayo, Peter Trumper, Ronald M. Pike—2nd ed.
        p.   cm.

    Includes bibliographical references.
    ISBN 0-471-24909-2 (pbk.:alk. paper)
    I. Chemistry, Organic–Laboratory manuals.   I. Trumper Peter K., 1955   II. Pike, Ronald M.   III. Title.

QD261 .M383 2001
547'0078—dc21   2001017672

ISBN 0-471-24909-2

10 9 8 7 6 5 4 3 2 1

Preface

In the Preface to the first edition of *Microscale Techniques for the Organic Laboratory* written in late 1990, we summed up the first decade and made predictions for the second as follows:

> The widespread application of microscale techniques to the introductory instructional laboratories began in the 1980s.[1] Since then the pace of microscale conversion has accelerated with the effects of this acceleration being experienced throughout chemical education and elsewhere. Microscale programs have been shown to be clearly superior in acquainting students with the operation of the chemical laboratory, but perhaps more importantly, they are also having a significantly positive impact on laboratory morale. At a recent seminar it was stated that 'the development of the microscale concept in the environs of the introductory organic laboratory is perhaps the most significant single advancement in chemical education in the last half century.'[2] Beyond the pedagogic benefits of miniaturization, the attendant improvements to safety, air quality, and waste generation can only be described as spectacular.[3] This environmentally sound style of laboratory instruction unquestionably will become a preferred educational pathway in the teaching of chemistry as the twentieth century draws to a close and the twenty-first century begins.

We are more than pleased to find a decade later that those comments were correct and the predictions to a great extent have come true. One good index of the widespread acceptance of this approach is that at latest count there are now at least seventeen microscale organic laboratory texts available from which the instructor may choose.

The decades of the 1980s and the 1990s did see a great deal of active reorganization of the instructional laboratories. Microscale techniques have played a major role in this transformation and not just in the introductory organic instructional laboratory. The first edition of *Microscale Techniques for the Organic Laboratory* (MTOL1) was aimed to provide the instructor with the maximum flexibility in the choice of organic chemistry to be included in his or her

---

[1]D. W. Mayo, R. M. Pike, and S. S. Butcher, *Microscale Organic Laboratory Preliminary 1*, John Wiley & Sons, Inc., 1985.

[2]Terence C. Morrill (RIT), Department of Chemistry, Merrimack College, April, 1989.

[3]D. W. Mayo, S. S. Butcher, R. M. Pike, C. M. Foote, J. R. Hotham, and D. S. Page, *J. Chem. Educ.* **1985,** *62,* 149.

personally designed microscale laboratory or the microscale component of a mixed-scale laboratory. MTOL1 appears to have successfully fulfilled these expectations. Interestingly, while there has been a veritable explosion of microscale-oriented laboratory texts, MTOL1 still stands alone as the sole instructional text available which is oriented primarily toward teaching the techniques of the microscale organic for the introductory laboratory. This text has also supplemented an additional educational void, that of microscale instruction for the research laboratory. Thus, MTOL1 has also found its way onto the shelves of numerous research laboratories around the country.

The text is designed to give a working view of the manipulative techniques and theory that one must master when undertaking experimental chemistry at the microscale level. The areas discussed are principally techniques that are used in the preparation, isolation, purification, and characterization of organic reaction products and naturally occurring materials. As in the first edition, we have included a selected set of experimental details in a few cases where relatively complex techniques are involved. In these sections this type of discussion greatly helps to underline just how the technique operates in the laboratory. Generally, the techniques can be easily adapted to local experimental apparatus although a few of the procedures, and apparatus used, do involve systems uniquely associated with the development of our particular program. In this regard, several examples of fractional distillation which employ bottom-driven spinning band distillation columns are given. One would be hard-pressed to carry out distillations with the required resolutions utilizing conventional packed columns. The preparative gas chromatography technique described utilizes a unique collection device, and the recrystallization techniques include a modified Teflon Craig head. These modifications make the manipulations significantly easier to perform, but are not essential to complete the operation.

Over the intervening years a number of the procedures that evolved during the development of the initial program have undergone modest or even significant modifications and refinements. The time has come to update MTOL1, and there are, therefore, extensive changes and improvements incorporated in this edition. A major change in editorial style has been the transfer of considerable theoretical source material and reference spectra to our Wiley website, www.wiley.com/college/mayo. This use of the website is similar to that found in the fourth edition of the original microscale laboratory text, *Microscale Organic Laboratory* (4th ed., John Wiley & Sons © 2000). The use of the website has allowed an expansion of the hard copy coverage into other areas. For example, *spectroscopy*, which occupied two chapters in the first edition, now includes two additional chapters involving introductory discussions of the application of ultraviolet and mass spectral data to the microscale organic laboratory. This expansion of the spectroscopic discussion became possible when large sections of infrared theory and reference spectra were moved to the Wiley website. As companion to the spectroscopic chapter, it has also been possible to include a new chapter that addresses the classic identification of organic compounds by qualitative organic chemistry.

As in the past, we would like to remember the pioneering efforts made in the development of instructional microscale programs and techniques by F. Emich and F. Pregl in Austria; N. D. Cheronis (who first defined 100 mg of starting substrate in an organic reaction as a microscale transformation), L. Craig, R. C. Fuson, E. H. Huntress, T. S. Ma, A. A. Morton, F. L. Schneider, and

R. L. Shriner in the United States; and J. T. Stock in both England and the United States. These educators laid the foundation on which we were able to fashion much of the current introductory program.

We continue to applaud the widespread development of affordable glassware for use in the microscale instructional laboratories. We are particularly pleased to note that in our experience the particular style of equipment (capseal connectors) which we developed for this program at Bowdoin College has accomplished an outstanding record of survival on the battleground of the sophomore laboratory bench. Much of the credit for the granite-like character of this equipment goes to J. Ryan and Larry Riley of the ACE Glass Company. We note with sadness the passing of Larry Riley since the first edition. His talent is sorely missed. A number of contributors have played long-term roles in the successful evolution of the microscale organic laboratory program, and we are happy to acknowledge them: Janet Hotham, Judy Foster, Lauren Bartlett, Robert Stevens, and Samuel Butcher all have made vital contributions along the way. We wish to add our appreciation of the support and encouragement that Jennifer Yee, our Associate Editor at Wiley has given us during the various stages of manuscript preparation.

To further enhance the dissemination of information and training for instructors in the area of microscale organic laboratory techniques, we note that a week-long Microscale Organic Chemistry workshop is now offered each summer at the National Microscale Chemistry Center located at Merrimack College, North Andover, MA 01845.

We are particularly indebted to our colleague Elizabeth Stemmler, the resident mass spectrometrist at Bowdoin College. Her contribution of the introductory discussion on the *Application of Mass Spectrometry to Organic Chemistry* launches the microscale laboratory into yet another area of spectroscopy that promises to add even further excitement to the introductory laboratory experience.

Finally, we are more than pleased to acknowledge the unique contributions of Paulette Messier, Laboratory Instructor at Bowdoin. Paulette is rapidly closing in on two decades of continuous laboratory instruction at the microscale level, a record of experience in microscale unmatched anywhere in the world of chemical education. Paulette, more than any one person has made this program a success in the trenches between the lab benches where it really counts. The thousands of students who have dealt directly with her, and gained her respect, are a tribute to Paulette's quiet, confident way of instilling enthusiasm and excitement into the microscale experience. Paulette Messier will always be indelibly linked to the Microscale Organic Laboratory at Bowdoin College.

As we stated earlier "the success or failure of the bench research chemist, to a very large extent, depends on an individual's ability to engage a vast array of laboratory skills while bringing about the solution to challenging experimental problems." We hoped that *Microscale Techniques for the Organic Laboratory* would help to ease the way to these solutions. We are pleased that our efforts have met that challenge. We hope our latest labors will aid a new generation of students down the exciting pathway of microscale organic chemistry.

December 2000
Brunswick, Maine

DANA W. MAYO
RONALD M. PIKE
PETER K. TRUMPER

Contents

CHAPTER 6

1

Introduction

You are about to embark on a challenging adventure—*the microscale organic chemistry laboratory!*

If you are a student assigned *Microscale Techniques for the Organic Laboratory* as a supplemental text in your introductory organic chemistry laboratory course, your experience is going to be quite different from that of the conventional manner in which this laboratory was taught for over a century. During the past two decades a large number of chemistry and pharmacology departments have found that the introductory organic laboratory course can be much more effectively taught at the microscale level. Thus, you will be learning the experimental side of organic chemistry on a greatly reduced scale than conventionally employed. Although you will be working with very small amounts of materials, you will, however, have the potential to observe and learn nearly as much experimental organic chemistry in one year as many of your predecessors or peers did or will do in approximately two years of conventional-scale laboratory instruction. You will have a very good chance of finding this educational experience an exciting and interesting one. While we cannot guarantee it for you individually, the majority of students who went through the program during its development nearly two decades ago found the microscale organic laboratory to be a surprisingly pleasant adventure.

This particular microscale organic laboratory textbook is centered on helping you develop skills in microscale organic laboratory techniques. It was developed to satisfy the needs of two different audiences. It is primarily designed to aid instructors and students involved in the instructional introductory organic laboratory, as it allows broad flexibility in developing local experiments and sequences that are not tied to a prescribed set of quantities. Second, for those readers working in the research environment at the advanced undergraduate or graduate level or engaged in industrial research, this text can provide a foundation from which it is possible to develop a solid expertise in microscale techniques. If you are unfamiliar with microscale organic chemistry, you are about to undergo a real shift of gears, as working at this level is, in reality, substantially different both in theory and practice from carrying out the conventional operations in the organic laboratory on multigram quantities of materials.

During the last decade, the experimental side of academic and industrial organic chemistry has moved ever closer to the microscale level. This conversion started in earnest over 20 years ago but has recently been spurred on by the rapidly accelerating cost of chemical waste disposal. It cannot be overemphasized that working with very small amounts of materials will allow you to learn, experience, and accomplish more organic chemistry in the long run.

First, we want to acquaint you with the organization and contents of the second edition of *Microscale Techniques for the Organic Laboratory* (MTOL2). We will then give you a few words of advice, which, if they are heeded, will allow you to avoid many of the sand traps you could land in as you develop microscale laboratory techniques. Finally, we wax philosophical and attempt to describe what we think you should derive from this experience.

Following this brief introduction, the second chapter is concerned with safety in the laboratory. This chapter supplies information that will allow you to estimate your maximum possible exposure to volatile chemicals used in the microscale laboratory. Chapter 2 also discusses general safety protocol for the laboratory. It is vitally important that you become familiar with the details of the material contained in this chapter; your health and safety depend on this knowledge.

The next three chapters, the heart of this book, are concerned primarily with the development of microscale experimental techniques. Chapter 3 describes in detail the experimental glassware employed in the microscale organic chemistry laboratory: the logic behind its construction, tips on its usage, the common arrangements of equipment, and various other laboratory manipulations, including techniques for transferring microquantities of materials, particularly liquids. Suggestions for the organization of your laboratory notebook are presented at the end of this chapter.

Chapter 4 deals with equipment and techniques for determining, on microscale samples, a number of conventional physical properties that are used in describing these organic substances. Chapter 5 is divided into nine technique sections. The detailed discussions of the techniques developed in this chapter and its associated website (🐚) establish the major experimental operations that are utilized in the microscale organic laboratory.

Chapters 6 through 9 develop the characterization of organic materials at the microscale level by spectroscopic techniques. Chapter 6 presents a brief discussion of the interpretation of infrared (IR) group frequencies. An introduction to the theory of this technique and the instrumentation involved in these studies is available via the MTOL2 website (🐚). The expanded discussions of Chapter 7 focus on an introduction to the latest developments in nuclear magnetic resonance (NMR) spectroscopy and its application to the elucidation of molecular structure.

Chapter 8, the first of three new chapters in MTOL2, contains a brief introductory discussion of the theory of ultraviolet-visible (UV–vis) spectroscopy and its application via a number of illustrated empirical examples. The new Chapter 9 presents a brief introduction to the theory, experimental techniques, and applications of mass spectrometry to organic chemistry.

The new Chapter 10 develops the characterization of organic materials at the microscale level by the use of classical organic reactions to form solid derivatives. Tables of derivative data for use in compound identification by these techniques are discussed and are found on the MTOL2 website (🌐).

In addition, we have introduced in this edition toward the end of most chapters and also after each of the major technique discussions in Chapter 5, and the experimental examples given in that chapter, a series of questions and problems designed to enhance and focus reader understanding of the experimental procedures in a particular laboratory technique or example. Finally, a bibliography consisting of a list of literature references is also found at the end of each chapter and following each technique discussed in Chapter 5. The bibliography section is a very important part of the text. Because the discussion of the microscale techniques involved is necessarily an introduction to these operations, we hope that you will take time to read and expand your knowledge about the particular technique being studied. You may, in fact, find that some of these references become assigned reading.

A prompt (➡) in the text indicates that experimental apparatus involved with that stage of the experiment are shown in the margin. Important comments are italicized in the text, and *Warnings* and *Cautions* are given in boxes and also indicated in the margins.

New to this edition is our accompanying website, www.wiley.com/college/MTOL2. To streamline our treatment of the subject we have moved a considerable quantity of basic reference material from the text and placed it in easily accessible form on our website. An icon (🌐) lets you know that supplemental material is available on the website. We hope this change will make the more important aspects of the basic text easier to access and speed your laboratory work along.

GENERAL RULES FOR THE MICROSCALE LABORATORY

1. *Study the experiment before you come to lab.* This rule is a historical plea from all laboratory instructors. In the microscale laboratory it takes on a more important meaning. You will not survive if you do not prepare ahead of time. In microscale experiments, operations happen much more quickly than in the macroscale laboratory. Your laboratory time will be overflowing with many more events. If you are not familiar with the sequences you are to follow, you will be in deep trouble. Although the techniques employed at the microscale level are not particularly difficult to acquire, they do demand a significant amount of attention. For you to reach a successful and happy conclusion, you cannot afford to have the focus of your concentration broken by having to constantly refer to the text during the experiment. Disaster is ever-present for the unprepared.

2. *ALWAYS work with clean equipment.* You must take the time to scrupulously clean your equipment before you start any experiment. Contaminated glassware will ultimately cost you additional time, and you will experience the frustration of inconsistent results and lower yields. Dirty equipment is the primary cause of reaction failure at the microscale level.

3. *CAREFULLY measure the quantities of materials to be used in the experiments.* A little extra time at the beginning of the laboratory can speed you on your way at the end of the session. A great deal of time has been spent optimizing the conditions employed in these experiments in order to maximize yields. Many organic reactions are very sensitive to the relative quantities of substrate (the material on which the reaction is taking place) and reagent (the reactive substance or substances that bring about the change in the substrate). After equipment contamination, the second-largest cause of failed reactions is attempting to run a reaction with incorrect quantities of the reactants present. Do not be hurried or careless at the balance.

4. *Clean means DRY.* Water or cleaning solution can be as detrimental to the success of a reaction as dirt or sludge in the system. You often will be working with very small quantities of moisture-sensitive reagents. The glass surface areas with which these reagents come in contact, however, are relatively large. A slightly damp piece of glassware can rapidly deactivate a critical reagent and result in reaction failure. *This rule must be strictly followed.*

5. *ALWAYS work on a clean laboratory bench surface, preferably glass!*

6. *ALWAYS protect the reaction product* that you are working with from a disastrous spill by carrying out all solution or solvent transfers over a crystallizing dish.

7. *ALWAYS place reaction vials or flasks in a clean beaker* when standing them on the laboratory bench. Then, when a spill occurs the material is more likely to be contained in the beaker and less likely to be found on the laboratory bench or floor.

8. *NEVER use cork rings to support round-bottom flasks,* particularly if they contain liquids. You are inviting disaster to be a guest at your laboratory bench.

9. *ALWAYS think through the next step* you are going to perform *before* starting it. Once you have added the wrong reagent, it is back to square one.

10. *ALWAYS save everything* you have generated in an experiment until it is successfully completed. You can retrieve a mislabeled chromatographic fraction from your locker, but not from the waste container!

THE ORGANIC CHEMISTRY LABORATORY

The confidence gained by mastering the microscale techniques described here will pay big dividends as you progress into modern-day experimental chemistry. The organic laboratory has had a reputation of being smelly, long, tedious, and pockmarked with fires and explosions; but present-day organic chemistry is undergoing a revolution at the laboratory bench. New techniques are sweeping away many of the old complaints, as an increasing fraction of industrial and academic research is being carried out at the microscale level.

This book allows the interested participant to rapidly develop the skills needed to slice more deeply into organic chemistry than ever before. The attendant benefits are greater confidence and independence in acquired laboratory techniques. The happy result is that in a microscale-based organic chemistry laboratory, you are more likely to have a satisfying encounter with the experimental side of this fascinating field of knowledge.

2

Safety

Research laboratories vary widely with respect to facilities and support given to safety. Large laboratories may have several hundred chemists and an extensive network of co-workers, supervisors, safety officers, and hazardous-waste managers. They also, according to governmental regulations, have an extensive set of safety procedures and detailed practices for the storage and disposal of hazardous wastes. In small laboratories, the individual chemist may have to take care of all these aspects of safety. Some laboratories may routinely deal with very hazardous materials and may run all reactions in hoods. Others may deal mainly with relatively innocuous compounds and have very limited hood facilities.

Our approach is to raise some questions to think about and to suggest places to look for further information. In this chapter, we do not present a large list of safety precautions for use in all situations; rather, we present a list of very basic precautionary measures. A bibliography at the end of the chapter offers a list of selected references. *We urge you to consult these references concerning specific safety regulations.* Many laboratories may have safety guidelines that will supercede this very cursory treatment. This chapter is no more than a starting point.

MAKING THE LABORATORY A SAFER PLACE

Murphy's law states in brief, "If anything can go wrong, it will." Although it is often taken to be a silly law, it is not. Murphy's law means that if sparking switches are present in areas that contain flammable vapors, sooner or later there will be a fire. If the glass container can move to the edge of the shelf as items are moved around or because the building vibrates, at some time it will come crashing to the floor. If the pipet can become contaminated, then the mouth pipetter will eventually ingest a contaminant.

We cannot revoke Murphy's law, but we can do a lot to minimize the damage. We can reduce the incidence of sparks and flames and flammable vapors. We can make sure that if the accident does occur, we have the means to contain the damage and to take care of any injuries that result. All of this means thinking about the laboratory environment. Does your laboratory have or enforce regulations related to important items such as eye, face, and foot protection, safety clothing, respiratory equipment, first aid supplies, fire equipment, spill kits, hoods, and compliance regulations? *Think ahead* about what could go wrong and then *plan* and *prepare* to minimize the chance of an accident and be prepared to respond when one does occur.

NATURE OF HAZARDS

The chemistry laboratory presents a wide assortment of risks. These risks are outlined briefly here so that you can begin to think about the steps necessary to make the laboratory safer:

1. *Physical hazards.* Injuries resulting from flames, explosions, and equipment (cuts from glass, electrical shock from faulty instrumentation, or improper use of instruments).
2. *External exposure to chemicals.* Injuries to skin and eyes resulting from contact with chemicals that have spilled, splashed, or been left on the bench top or on equipment.
3. *Internal exposure.* Longer term (usually) health effects resulting from breathing hazardous vapors or ingesting chemicals.

REDUCTION OF RISKS

Many things can be done to reduce risks. The rules below may be absolute in some laboratories. In others, the nature of the materials and apparatus used may justify the relaxation of some of these rules or the addition of others.

1. *Stick to the procedures described by your supervisor.* This attention to detail is particularly important for the chemist with limited experience. In other cases, variation of the reagents and techniques may be part of the work.
2. *Wear approved safety goggles.* We can often recover quickly from injuries affecting only a few square millimeters on our bodies, unless that area happens to be in our eyes. Larger industrial laboratories often require that laboratory work clothes and safety shoes be worn. Wear them, if requested.
3. *Do not put anything in your mouth under any circumstances while in the laboratory.* This includes food, drinks, chemicals, and pipets. There are countless ways that surfaces can become contaminated in the laboratory. Since there are substances that must never be pipetted by mouth, one must get into the habit of *never* mouth–pipetting anything.

4. *Be cautious with flames and flammable solvents.* Remember that the flame at one end of the bench can ignite the flammable liquid at the other end in the event of a spill or improper disposal. Flames must never be used when certain liquids are present in the laboratory, and flames must always be used with care. Check the *fire diamond* hazard symbol, if available.

5. *Be sure that you have the proper chemicals for your reaction.* Check labels carefully, and return unused chemicals to the proper place for storage. Be sure to replace caps on containers immediately after use. An open container is an invitation for a spill. Furthermore, some reagents are very sensitive to moisture, and may decompose if left open.

6. *Minimize the loss of chemicals to air or water and dispose of waste properly.* Some water-soluble materials may be safely disposed of in the water drains. Other wastes should go into special receptacles. Pay attention to the labels on these receptacles. Recent government regulations have placed stringent rules on industrial and academic laboratories for proper disposal of chemicals. *Severe penalties are levied on those who do not follow proper procedures.* We recommend that you consult general safety references nos. 3 and 4 at the end of the chapter.

7. *Minimize skin contact with any chemicals.* Use impermeable gloves when necessary, and promptly wash any chemical off your body. If you have to wash something off with water, use lots of it. Be sure that you know where the nearest water spray device is located.

Note

Do not use latex gloves. They are permeable to many chemicals, and some people are allergic to them.

8. *Do not inhale vapors from volatile materials.* Severe illness or internal injury can result.

9. *Tie back or confine long hair and loose items of clothing.* You do not want them falling into a reagent or getting near flames.

10. *Do not work alone.* Too many things can happen to a person working alone that might leave him or her unable to obtain assistance. As in swimming, the "buddy system" is safest.

11. *Exercise care in assembling glass and electrical apparatus.* All operations with glass, such as separating standard taper glassware, involve the risk that the glass may break and that lacerations or punctures may result. Seek help or advice with glassware, if necessary. Special containers should be provided for the disposal of broken glass. Electrical shock can occur in many ways. When making electrical connections, make sure that your hands, the laboratory bench, and the floor are all dry and that *you* do not complete an electrical path to ground. Be sure that electrical equipment is properly grounded and insulated.

12. *Report any injury or accident to the appropriate person.* Reporting injuries and accidents is important so that medical assistance can be obtained if necessary. It also allows others to be made aware of any safety problems; these problems may be correctable.

13. *Keep things clean.* Put unused apparatus away. Immediately wipe up or care for spills on the bench top or floor. This also pertains to the balance area and to where chemicals are dispensed.

14. *Never heat a closed system.* Always provide a vent to avoid an explosion. Provide a suitable trap for any toxic gases generated in a given reaction.

15. *Learn the correct use of gas cylinders.* Even a small gas cylinder can become a lethal bomb if not properly used.

16. *Attend safety programs.* Many industrial laboratories offer excellent seminars and lectures on a wide variety of safety topics. Pay careful attention to the advice and counsel of the safety officer.

17. *Above all, use your common sense.* Think before you act.

PRECAUTIONARY MEASURES

Know the location and operation of safety equipment in the laboratory. Locate the nearest

- Fire extinguisher
- Telephone
- Exit
- First aid kit
- Emergency shower
- Fire blanket
- Eye wash

Know where to call (have the numbers posted) for

- Fire
- Medical emergency
- Spill or accidental release of corrosive or toxic chemicals

Know where to go

- In case of injury
- To evacuate the building

THINKING ABOUT THE RISKS IN USING CHEMICALS

The smaller quantities used in the microscale laboratory carry with them a reduction in hazards caused by fires and explosions; hazards associated with skin contact are also reduced. However, care must be exercised when working with even the small quantities involved.

There is great potential for reducing the exposure to chemical vapors, but these reductions will be realized only if everyone in the laboratory is careful. One characteristic of vapors emitted outside hoods is that they mix rapidly throughout the lab and will quickly reach the person on the other side of the room. In some laboratories, the majority of reactions may be carried out in hoods. When reactions are carried out in the open laboratory, each experimenter becomes a polluter whose emissions affect nearby people the most, but these emissions become added to the laboratory air and to the burden each of us must bear.

The concentration of vapor in the general laboratory air space depends on the vapor pressure of the liquids, the area of the solid or liquid exposed, the nature of air currents near the sources, and the ventilation characteristics of the laboratory. One factor over which each individual has control is evaporation, which can be reduced by the following practices:

* Certain liquids must remain in hoods.
* Reagent bottles must be recapped when not in use.
* Spills must be quickly cleaned up and the waste discarded.

Chemicals must be properly stored when not in use. Some balance must be struck between the convenience of having the compound in the laboratory where you can easily put your hands on it and the safety of having the compound in a properly ventilated and fire-safe storage room. Policies for storing chemicals will vary from place to place. There are limits to the amounts of flammable liquids that should be stored in glass containers, and fire-resistant cabinets must be used for storage of large amounts of flammable liquids. Chemicals that react with one another should not be stored in close proximity. There are plans for sorting chemicals by general reactivity classes in storerooms; for instance, Flinn Scientific Company includes a description of a storage system in their (1997) chemical catalog.

DISPOSAL OF CHEMICALS

Chemicals must also be segregated into categories for disposal. The categories used will depend on the disposal service available and upon federal, state, and local regulations. For example, some organic wastes are readily incinerated, while those containing chlorine may require much more costly treatment. Other wastes may have to be buried. For safety and economic reasons, it is important to place waste material in the appropriate container. In today's world, it often costs more to dispose of a chemical than to purchase it in the first place! The economic impact of waste generation and disposal is gigantic. Do you realize that the chemical industry in the United States releases more than 3 billion tons of chemical waste each year? Then $150 billion is spent per year in waste treatment, control, and disposal costs!

It is our obligation as chemists to decrease the impact that hazardous chemicals have on our environment. A movement is currently underway (referred to as "Green Chemistry" or "Benign by Design") to accomplish this goal by focusing on the design, manufacture, and use of chemicals and chemical processes that have little or no pollution potential or environmental risk.

MATERIAL SAFETY DATA SHEETS

Although risks are associated with the use of most chemicals, the magnitudes of these risks vary greatly. A short description of the risks is provided by a Material Safety Data Sheet, commonly referred to as an MSDS. These sheets are normally provided by the manufacturer or vendor of the chemical, and users are required to keep on file the MSDS of each material stored or used. Aldrich Chemical Company has available over 70,000 complete, printable data sheets on CD-ROM; the information is updated quarterly.

As an example, the MSDS for acetone is shown here. This sheet is provided by the J. T. Baker Chemical Company. Sheets from other sources will be very similar. Much of the information on these sheets is self-explanatory, but let's review the major sections of the acetone example.

Section I provides identification numbers and codes for the compound and includes a summary of the risks associated with the use of acetone. Because these sheets are available for many thousands of compounds and mixtures, there must be a means of unambiguously identifying the substance. A standard reference number for chemists is the Chemical Abstracts Service Number (CAS No.).

A quick review of the degree of risks is given by the numerical scale under Precautionary Labeling. This particular scale is a proprietary scale that ranges from 0 (very little or nonexistent risk) to 4 (extremely high risk). The National Fire Protection Association (NFPA) uses a similar scale, but the risks considered are different. Other systems may use different scales, and there are some that represent low risks by the highest number! Be sure that you understand the scale being used. Perhaps some day one scale will become standard.

Section II covers risks from mixtures. Because a mixture is not considered here, the section is empty. Selected physical data are described in Section III. Section IV contains fire and explosion data, including a description of the toxic gases produced when acetone is exposed to a fire. The MSDSs are routinely made available to fire departments that may be faced with fighting a fire in a building where large amounts of chemicals are stored.

Health hazards are described in Section V. The entries of most significance for evaluating risks from vapors are the Threshold Limit Value (or TLV) and the Short-Term Exposure Limit (STEL). The TLV is a term used by the American Conference of Govermental Industrial Hygienists (ACGIH). This organization examines the toxicity literature for a compound and establishes the TLV. This standard is designed to protect the health of workers exposed to the vapor 8 hours a day, five days a week. The Occupational Safety and Health Administration (OSHA) adopts a value to protect the safety of workplaces in the United States. Their value is termed the Time-Weighted Average (TWA) and in many cases is numerically equal to the TLV. The STEL is a value not to be exceeded for even a 15-minute averaging time. TLV, TWA, and STEL values for many chemicals are summarized in a small handbook available from the ACGIH (1990); they are also collected in the *CRC Handbook of Chemistry and Physics.*

The toxicity of acetone is also described in terms of the toxic oral dose. In this case, the LD_{50} is the dose that will cause the death of 50% of the mice or rats given that dose. The dose is expressed as milligrams of chemical per kilogram of body weight of the subject animal. The figures for small animals are often used to estimate the effects on humans. If, for example, we used the mouse figure of 1297 mg/kg and applied it to a 60-kg chemist, a dose of 77,820 mg ($\sim$98.5 mL) would kill 50% of the subjects receiving that dose. As a further example, chloroform has an LD_{50} of 80 mg/kg. For our 60-kg chemist, a dose of 4800 mg ($\sim$3 mL) would be fatal for 50% of these cases. The effects of exposure of skin to the liquid and vapor are also described.

J. T. BAKER CHEMICAL CO. 222 RED SCHOOL LANE, PHILLIPSBURG, NJ 08865
M A T E R I A L   S A F E T Y   D A T A   S H E E T
24-HOUR EMERGENCY TELEPHONE — (201) 859-2151
CHEMTREC # (800) 424-9300 — NATIONAL RESPONSE CENTER # (800) 424-8802

A0446 –01   ACETONE   PAGE: 1
EFFECTIVE: 10/11/85   ISSUED: 01/23/86

SECTION I – PRODUCT IDENTIFICATION

PRODUCT NAME: ACETONE
FORMULA: (CH3)2CO
FORMULA WT: 58.08
CAS NO.:   00067-64-1
NIOSH/RTECS NO.:   AL3150000
COMMON SYNONYMS: DIMETHYL KETONE; METHYL KETONE; 2-PROPANONE
PRODUCT CODES: 9010,9006,9002,9254,9009,9001,9004,5356,A134,9007,9005,9008

PRECAUTIONARY LABELLING

BAKER SAF-T-DATA(TM) SYSTEM

          HEALTH   – 1
          FLAMMABILITY   – 3 (FLAMMABLE)
          REACTIVITY   – 2
          CONTACT   – 1

LABORATORY PROTECTIVE EQUIPMENT

SAFETY GLASSES; LAB COAT; VENT HOOD; PROPER GLOVES; CLASS B EXTINGUISHER

PRECAUTIONARY LABEL STATEMENTS

                        DANGER
                  EXTREMELY FLAMMABLE
            HARMFUL IF SWALLOWED OR INHALED
                   CAUSES IRRITATION
KEEP AWAY FROM HEAT, SPARKS, FLAME. AVOID CONTACT WITH EYES, SKIN, CLOTHING.
AVOID BREATHING VAPOR. KEEP IN TIGHTLY CLOSED CONTAINER. USE WITH ADEQUATE
VENTILATION. WASH THOROUGHLY AFTER HANDLING. IN CASE OF FIRE, USE WATER SPRAY,
ALCOHOL FOAM, DRY CHEMICAL, OR CARBON DIOXIDE. FLUSH SPILL AREA WITH WATER
SPRAY.

SECTION II – HAZARDOUS COMPONENTS

COMPONENT	%	CAS NO.
ACETONE	90-100	67-64-1

SECTION III – PHYSICAL DATA

BOILING POINT:	56 C (133 F)	VAPOR PRESSURE(MM HG):	181
MELTING POINT:	−95 C (−139 F)	VAPOR DENSITY(AIR=1):	2
SPECIFIC GRAVITY:	0.79	EVAPORATION RATE:	5.6
(H2O=1)		(BUTYL ACETATE=1)	

SOLUBILITY(H2O):   COMPLETE (IN ALL PROPORTIONS) % VOLATILES BY VOLUME: 100

APPEARANCE & ODOR: CLEAR, COLORLESS LIQUID WITH FRAGRANT SWEET ODOR.

SECTION IV – FIRE AND EXPLOSION HAZARD DATA

FLASH POINT: −18 C (0 F)   NFPA 704M RATING: 1-3-0

FLAMMABLE LIMITS: UPPER – 13 %   LOWER –   2 %

FIRE EXTINGUISHING MEDIA
    USE ALCOHOL FOAM, DRY CHEMICAL OR CARBON DIOXIDE.
    (WATER MAY BE INEFFECTIVE.)

SPECIAL FIRE-FIGHTING PROCEDURES
    FIREFIGHTERS SHOULD WEAR PROPER PROTECTIVE EQUIPMENT AND SELF-CONTAINED
    (POSITIVE PRESSURE IF AVAILABLE) BREATHING APPARATUS WITH FULL FACEPIECE.
    MOVE EXPOSED CONTAINERS FROM FIRE AREA IF IT CAN BE DONE WITHOUT RISK.
    USE WATER TO KEEP FIRE-EXPOSED CONTAINERS COOL.

UNUSUAL FIRE & EXPLOSION HAZARDS
    VAPORS MAY FLOW ALONG SURFACES TO DISTANT IGNITION SOURCES AND FLASH BACK.
    CLOSED CONTAINERS EXPOSED TO HEAT MAY EXPLODE. CONTACT WITH STRONG
    OXIDIZERS MAY CAUSE FIRE.

SECTION V – HEALTH HAZARD DATA

THRESHOLD LIMIT VALUE (TLV/TWA):   1780 MG/M3 (750 PPM)

SHORT-TERM EXPOSURE LIMIT (STEL): 2375 MG/M3 (1000 PPM)

TOXICITY: LD50 (ORAL-RAT)(MG/KG)   – 9750
              LD50 (IPR-MOUSE)(G/KG)   – 1297

EFFECTS OF OVEREXPOSURE
>   CONTACT WITH SKIN HAS A DEFATTING EFFECT, CAUSING DRYING AND IRRITATION.
>   OVEREXPOSURE TO VAPORS MAY CAUSE IRRITATION OF MUCOUS MEMBRANES, DRYNESS
>   OF MOUTH AND THROAT, HEADACHE, NAUSEA AND DIZZINESS.

EMERGENCY AND FIRST AID PROCEDURES
>   CALL A PHYSICIAN.
>   IF SWALLOWED, IF CONSCIOUS, IMMEDIATELY INDUCE VOMITING.
>   IF INHALED, REMOVE TO FRESH AIR. IF NOT BREATHING, GIVE ARTIFICIAL
>   RESPIRATION. IF BREATHING IS DIFFICULT, GIVE OXYGEN.
>   IN CASE OF CONTACT, IMMEDIATELY FLUSH EYES WITH PLENTY OF WATER FOR AT
>   LEAST 15 MINUTES. FLUSH SKIN WITH WATER.

SECTION VI – REACTIVITY DATA

STABILITY: STABLE   HAZARDOUS POLYMERIZATION: WILL NOT OCCUR

CONDITIONS TO AVOID: HEAT, FLAME, SOURCES OF IGNITION

INCOMPATIBLES: SULFURIC ACID, NITRIC ACID, STRONG OXIDIZING AGENTS

SECTION VII – SPILL AND DISPOSAL PROCEDURES

STEPS TO BE TAKEN IN THE EVENT OF A SPILL OR DISCHARGE
>   WEAR SUITABLE PROTECTIVE CLOTHING. SHUT OFF IGNITION SOURCES; NO FLARES,
>   SMOKING, OR FLAMES IN AREA. STOP LEAK IF YOU CAN DO SO WITHOUT RISK. USE
>   WATER SPRAY TO REDUCE VAPORS. TAKE UP WITH SAND OR OTHER NON-COMBUSTIBLE
>   ABSORBENT MATERIAL AND PLACE INTO CONTAINER FOR LATER DISPOSAL. FLUSH
>   AREA WITH WATER.

>   J. T. BAKER SOLUSORB(R) SOLVENT ADSORBENT IS RECOMMENDED
>   FOR SPILLS OF THIS PRODUCT.

DISPOSAL PROCEDURE
>   DISPOSE IN ACCORDANCE WITH ALL APPLICABLE FEDERAL, STATE, AND LOCAL
>   ENVIRONMENTAL REGULATIONS.

EPA HAZARDOUS WASTE NUMBER: U002 (TOXIC WASTE)

SECTION VIII – PROTECTIVE EQUIPMENT

VENTILATION: USE GENERAL OR LOCAL EXHAUST VENTILATION TO MEET
                                TLV REQUIREMENTS.

RESPIRATORY PROTECTION: RESPIRATORY PROTECTION REQUIRED IF AIRBORNE
                                CONCENTRATION EXCEEDS TLV. AT CONCENTRATIONS UP
                                TO 5000 PPM, A GAS MASK WITH ORGANIC VAPOR
                                CANNISTER IS RECOMMENDED. ABOVE THIS LEVEL, A
                                SELF-CONTAINED BREATHING APPARATUS WITH FULL FACE
                                SHIELD IS ADVISED.

EYE/SKIN PROTECTION: SAFETY GLASSES WITH SIDESHIELDS, POLYVINYL ACETATE
                                GLOVES ARE RECOMMENDED.

SECTION IX – STORAGE AND HANDLING PRECAUTIONS

SAF-T-DATA(TM) STORAGE COLOR CODE: RED

SPECIAL PRECAUTIONS
>   BOND AND GROUND CONTAINERS WHEN TRANSFERRING LIQUID. KEEP CONTAINER
>   TIGHTLY CLOSED. STORE IN A COOL, DRY, WELL-VENTILATED, FLAMMABLE LIQUID
>   STORAGE AREA.

SECTION X – TRANSPORTATION DATA AND ADDITIONAL INFORMATION

DOMESTIC (D.O.T.)

PROPER SHIPPING NAME	ACETONE
HAZARD CLASS	FLAMMABLE LIQUID
UN/NA	UN1090
LABELS	FLAMMABLE LIQUID

INTERNATIONAL (I.M.O.)

PROPER SHIPPING NAME	ACETONE
HAZARD CLASS	3.1
UN/NA	UN1090
LABELS	FLAMMABLE LIQUID

(TM) AND (R) DESIGNATE TRADEMARKS.
N/A = NOT APPLICABLE OR NOT AVAILABLE

THE INFORMATION PUBLISHED IN THIS MATERIAL SAFETY DATA SHEET HAS BEEN COMPILED
FROM OUR EXPERIENCE AND DATA PRESENTED IN VARIOUS TECHNICAL PUBLICATIONS. IT IS
THE USER'S RESPONSIBILITY TO DETERMINE THE SUITABILITY OF THIS INFORMATION FOR
THE ADOPTION OF NECESSARY SAFETY PRECAUTIONS. WE RESERVE THE RIGHT TO REVISE
MATERIAL SAFETY DATA SHEETS PERIODICALLY AS NEW INFORMATION BECOMES AVAILABLE.

Section VI describes the reactivity of acetone and the classes of compounds with which it should not come in contact. For example, sodium metal reacts violently with a number of substances (including water) and should not come in contact with them. Strong oxidizing agents (such as nitric acid) should not be mixed with organic compounds (among other things). The final sections (Sections VII–X) are self-explanatory.

ALTERNATE SOURCES OF INFORMATION

Similar information in a more compact form can be found in the *Merck Index* (Merck). This basic reference work provides information on the toxicity of many chemicals. It often refers one to the *NIOSH Pocket Guide to Chemical Hazards* (National Institute for Occupational Safety and Health). The *Merck Index* also supplies interesting information about the common uses of the chemicals listed, particularly related to the medical area. References to the chemical literature are also provided. The *CRC Handbook of Chemistry and Physics*, which is updated each year, contains a wide range of data (located in tables) in the area of health, safety, and environmental protection. It also includes directions for the handling and disposal of laboratory chemicals. Your laboratory should have a copy of this work. Most chemical supply houses now label their containers with data showing not only the usual package size, physical properties, and chemical formula, but also pictures or codes showing hazard information. Some include a pictogram (for example, see the newer Aldrich Chemical labels on their bottles). The J. T. Baker Company uses the Baker SAF-T-DATA System.

ESTIMATING RISKS FROM VAPORS

Other things (availability, suitability) being equal, one would, of course, choose the least toxic chemical for a given reaction. Some very toxic chemicals play very important roles in synthetic organic chemistry, and the toxicity of the chemicals in common use varies greatly. Bromine and benzene have TLVs of 0.7 and 30 mg/m^3, respectively, and are at the more toxic end of the spectrum of chemicals routinely used. Acetone has a TLV of 1780 mg/m^3. These representative figures do not mean that acetone is harmless or that bromine cannot be used. In general, one should exercise care at all times (make a habit of good laboratory practice) and should take special precautions when working with highly toxic materials.

The TLV provides a simple means to evaluate the relative risk of exposure to the vapor of any substance used in the laboratory. If the quantity of the material evaporated is represented by m (in milligrams/hour) and the TLV is expressed by L (milligrams per cubic meter), a measure of relative risk to the vapor is given by m/L. This quantity represents the volume of clean air required to dilute the emissions to the TLV. As an example, the emissions of 1 g of bromine and 10 g of acetone in one hour lead to the values of m/L of 1400 m^3/hour (h) for the bromine and 5.6 m^3/h for acetone. These numbers provide a direct handle on the *relative* risks from these two vapors. It is difficult to assess the absolute risk to these vapors without a lot of information about the ventilation characteristics of the laboratory. If these releases occur within a properly operated hood, the threat to the worker in the laboratory is probably very small. (However, consideration must be given to the hood exhaust.)

Exposure in the general laboratory environment can be assessed if we assume that the emissions are reasonably well mixed before they are inhaled and if we know something about the room ventilation rate. The ventilation rate of the room can be measured by a number of ways.[1] Given the ventilation rate, it might be safe to assume that only 30% of that air is available for diluting the emissions. (This accounts for imperfect mixing in the room.) The effective amount of air available for dilution can then be compared with the amount of air required to dilute the chemical to the TLV.

Let us continue our example. Suppose that the laboratory has a volume of 75 m^3 and an air exchange rate of 2 air changes per hour. This value means that $(75m^3)(2/h)(0.3) = 45$ m^3/h are available to dilute the pollutants. There may be enough margin for error to reduce the acetone concentration to a low level (5.6 m^3/h is required to reach the TLV), but use of bromine should be restricted to the hood. An assessment of the accumulative risk of several chemicals is obtained by adding the individual m/L $\left(\frac{mg/h}{mg/m^3}\right)$ values.

The m/L figures may also be used to assess the relative risk of performing the experiment outside a hood. Since m/L represents the volume of air for each student, this may be compared with the volume of air actually available for each student. If the ventilation rate for the entire laboratory is Q (in cubic meters per minute) for a section of n students meeting for t minutes, the volume for each student is kQt/n cubic meters. Here k is a mixing factor that allows for the fact that the ventilation air will not be perfectly mixed in the laboratory before it is exhausted. In a reasonable worst-case mixing situation a k value of 0.3 seems reasonable. Laboratories with modest ventilation rates supplied by 15–20 linear feet of hoods can be expected to provide 30–100 m^3 per student over a 3-h laboratory period if the hoods are working properly. Let us take the figure of 50 m^3 per student as an illustration. If the value of m/L for a compound (or a group of compounds in a reaction) is substantially less than 50 m^3, it may be safe to do that series of operations in the open laboratory. If m/L is comparable to or greater than 50 m^3, a number of options are available: (1) Steps using that compound may be restricted to a hood. (2) The instructional staff may satisfy themselves that much less than the assumed value is actually evaporated under conditions present in the laboratory. (3) The number of individual repetitions of this experiment may be reduced. The size of the laboratory section can be reduced or the experiment may be done in pairs or trios.

Conducting reactions in a hood does not automatically convey a stamp of safety. Hoods are designed to keep evaporating chemicals from entering the general laboratory space. For hoods to do their job, there must be an adequate flow of air into the hood, and this air flow must not be disturbed by turbulence at the hood face. A frequently used figure of merit for hood operation is the face velocity of 100 ft/min. This is an average velocity of air entering the hood opening. (Instruments for measuring this flow rate are available in the catalogs of major equipment suppliers. Prices range from less than $50 to several hundred dollars.) Even with a face velocity of 100 ft/min, vapors can be

---

[1]Butcher, S. S.; Mayo, D. W.; Hebert, S. M.; Pike, R. M., "Laboratory Air Quality, Part I"; *J. Chem. Educ.* **1985**, *62*, A238; and "Laboratory Air Quality, Part II" *J. Chem. Educ.* **1985**, *62*, A261.

drawn out of an improperly designed hood simply by people walking by the opening, or by drafts from open windows.

Hood performance should be checked at regular intervals. The face velocity will increase as the front hood opening is decreased. If an adequate face velocity cannot be maintained with a front opening height of 15 cm, use of the hood for carrying out reactions will be limited. A low face velocity may indicate that the fans and ductwork need cleaning, that the exhaust system leaks (if it operates under lower than ambient pressure), or that the supply of makeup air is not adequate. When the hood system is properly maintained, the height of the hood opening required to provide an adequate face velocity is often indicated with a sticker.

Hoods are often used for storage of volatile compounds. A danger in this practice is that the hood space can become quickly cluttered, making work in the hood difficult, and the air flow may be disturbed. Of course, hoods being used for storage must never be turned off.

CONCLUDING THOUGHTS

This brief chapter touches only a few of the important points concerning laboratory safety. The risk from vapor exposure is discussed in some detail, but other risks are treated briefly. Applications in some laboratories may involve reactions with a risk from radiation or infection or may involve compounds that are unstable with respect to explosion. The chemist must be aware of the potential risks and must be prepared to go to an appropriate and detailed source of information, as needed. The references cited here represent a small fraction of the safety data, texts, and journals available on this subject. It is highly recommended that the library and/or laboratory at your institution have at least this minimal selection. Of course, the selections should be kept up to date!

QUESTIONS

2-1. If your chemistry department has the MSDSs on CD-ROM, locate a chemical of your choice and print out the data. If the information is not available on CD-ROM, go to your stockroom and request a copy of the MSDS. Underline on the sheet the CAS No., solubility data, fire and explosion data, reactivity data, and what protective equipment is required when using this chemical. Does your laboratory meet the safety regulations required to use this chemical? Why or why not?

2-2. Think and describe what you would do in each of the following situations which could happen in your laboratory.

   (a) You are working at your station, and the 100-mL round-bottom flask in which you are running a reaction in ether solvent suddenly catches fire.

   (b) The person working across the laboratory bench from you allows hydrogen chloride gas to escape from his or her apparatus.

   (c) A reagent bottle is dropped, spilling concentrated sulfuric acid.

   (d) A hot solution "bumps," splashing your face.

2-3. You are working in the laboratory using 3.0 mL of benzene in an extraction procedure. An alternative to benzene is toluene. However, three times more toluene is required to perform the extraction. The isolation of the desired product from the extraction solution requires evaporation of the solvent (benzene or toluene). This takes 0.5 h to complete. Calculate the relative risks of using these two solvents. Which solvent would you use and why?

2-4. A laboratory has four hoods; each is 39 in. wide. When the hood door is open to a height of 8 in. and the hoods are operating, the average air velocity through the hood face is 170 ft/min.

    (a) Evaluate the total ventilation rate for this room, assuming that there are no other exhausts.

    (b) The laboratory is designed for use by 30 students. Evaluate the air available per student if the mixing factor is 0.3 and the experiments last for 3 h.

    (c) An experiment is considered in which each student would be required to evaporate 7 mL of methylene chloride (CH_2Cl_2). Estimate the average concentration of methylene chloride. Look up the TLV or the TWA for methylene chloride and consider how the evaporation might be performed.

2-5. An experiment is considered in which 1 mL of diethylamine would be used by each student. The ventilation rate for the laboratory is 5 m^3/min. Look up the TLV (or TWA) for diethylamine, $(C_2H_5)_2NH$. What restrictions might be placed on the laboratory to keep the average concentration, over a 3-h period, less than one-third of the TWA? Assume a mixing factor of 0.3.

GENERAL SAFETY REFERENCES

1. American Chemical Society, *Safety in Academic Chemical Laboratories*, 6th ed.; American Chemical Society: Washington, DC, 1995.

2. *Handbook of Laboratory Safety*, 4th ed.; Furr, A. K. Jr., Ed.; CRC Press: Boca Raton, FL, 1995.

3. Committee on Prudent Practices for Handling, Storage, and Disposal of Chemicals in Laboratories, National Research Council, *Prudent Practices for Handling Chemicals in Laboratories*; National Academy Press: Washington, DC, 1995.

4. Armour, M. A. *Hazardous Laboratory Chemicals Disposal Guide*, 2nd ed.; CRC Press: Boca Raton, FL, 1996.

5. *Working Safely with Chemicals in the Laboratory*, 2nd ed.; Gorman, C. E., Ed.; Genium: Schenectady, NY, 1995.

6. Hall, S. K. *Chemical Safety in the Laboratory*, CRC Press: Boca Raton, FL, 1994.

7. *The Sigma–Aldrich Library of Regulatory and Safety Data*; Lenga, R. E.; Votoupal, K. L., Eds.; Aldrich Chemical Co., Milwaukee, WI, 1992.

8. Young, J. A., *Improving Safety in the Chemical Laboratory: A Practical Guide*, 2nd ed.; Wiley: New York, 1991.

9. American Chemical Society, *Less Is Better (Laboratory Chemical Management for Waste Reduction)*, American Chemical Society: Washington, DC, 1993.

10. Verschueren, K., *Handbook of Environmental Data on Organic Chemicals*, 3rd ed.; Van Nostrand Reinhold: New York, NY, 1996.

11. Lewis, R. J. Sr., *Hazardous Chemicals Desk Reference*, 4th ed.: Van Nostrand Reinhold: New York, 1997.

12. *NIOSH Pocket Guide to Chemical Hazards*, National Institute for Occupational Safety and Health, U. S. Government Publication Office: Washington, DC, 1990. A diskette version is available from the Canadian Centre for Occupational Health and Safety. It can also be found on the Centre's CHEMSource CD-ROM.

13. *OSHA Regulated Hazardous Substances*, Vols. I and II, Noyes Data Corp.: Park Ridge, NJ, 1990.

14. Lund, G.; Sansone, E. B. "Safe Disposal of Highly Reactive Chemicals"; *J. Chem. Educ.* **1994**, *71*, 972

15. It is also recommended that one refer to the numerous articles on safety that appear regularly in the *Journal of Chemical Education* and *The Chemical Educator*.

BIBLIOGRAPHY

ACGIH. *Threshold Limit Values and Biological Exposure Indices.* Available from ACGIH, Kemper Woods Center, 1330 Kemper Meadow Drive, Cincinnati, OH 45240

Aldrich Catalog Handbook of Fine Chemicals, 1001 W. St. Paul Ave., Milwaukee, WI.

ANASTAS, P. T.; FARRIS, C. A., Eds., *"Benign by Design—Alternate Synthetic Design for Pollution Prevention,"* ACS Symposium Series 577, American Chemical Society: Washington, DC, 1994.

ANASTAS, P. T.; WILLIAMSON, T. C., Eds., *"Green Chemistry—Designing Chemistry for the Environment,"* ACS Symposium Series 626, American Chemical Society: Washington, DC, 1996.

BUTCHER, S. S.; MAYO, D. W.; HEBERT, S. M.; PIKE, R. M.," Laboratory Air Quality, Part I," *J. Chem. Educ.* **1985**, *62*, A238; "Laboratory Air Quality, Part II"; *J. Chem. Educ.* **1985**, *62*, A261.

Flinn Scientific Company, *Chemical Catalog/Reference Manual* (1997). Available from Flinn Scientific Co., P.O. Box 219, Batavia, IL 60510

Handbook of Chemistry and Physics, 78th ed.; Lide, D. R., Ed.; CRC Press: Boca Raton, FL, 1998.

The Merck Index, 12th ed.; Budavari, S., Ed.; Merck Research Laboratories Publications: Whitehouse Station, NJ, 1996.

MOLLINELLI, R. P.; Reale, M. J.; Freudenthal, R. I., *Material Data Safety Sheets*, Hill & Gernett: Boca Raton, FL, 1992.

3

Introduction to Microscale Organic Laboratory Equipment and Techniques

We begin this chapter with a description of the standard pieces of glassware that are generally employed in a microscale laboratory. Modern standard taper glassware is particularly convenient to use and gives the student a sense of the flavor of the research laboratory. It is not essential, however, for the experimental work in an instructional laboratory, and many courses use glassware with alternative connectors. We describe the standard taper glassware as just one example of microscale equipment that is available. The operations carried out in the laboratory will be very similar or identical if, for example, a plastic connector is used to assemble the experimental setup. We next consider a series of standard experimental apparatus setups that use this equipment, and present a short discussion of the role that they play in the laboratory. We end the chapter with a set of laws that govern how one operates in a microscale laboratory (the rules are a bit different than those for a macroscale laboratory) and a set of guidelines for recording your experimental data. The basic individual pieces of equipment are shown in Figures 3.1 to 3.7.

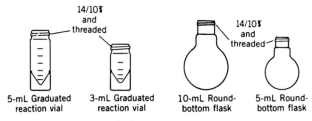

Figure 3.1 Reaction flasks.

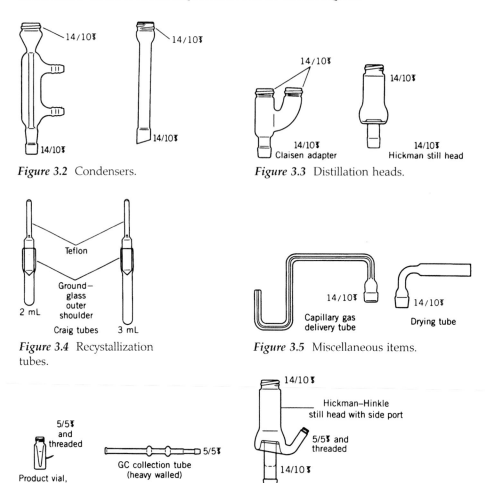

Figure 3.2 Condensers.

Figure 3.3 Distillation heads.

Figure 3.4 Recystallization tubes.

Figure 3.5 Miscellaneous items.

Figure 3.6 Gas chromatographic fraction collection items.

Figure 3.7 Hickman-Hinkle distillation column.

MICROGLASSWARE EQUIPMENT

Standard taper joints

Standard taper ground-glass joints are the common mechanism for assembling all conventional research equipment in the organic laboratory. The symbol ℑ is commonly used to indicate the presence of this type of connector. Normally, ℑ is either followed or preceded by #/#. The first # refers to the maximum inside diameter of a female (outer) joint or the maximum outside diameter of a male (inner) joint, measured in millimeters. The second number corresponds to the total length of the ground surface of the joint (Fig. 3.8). The advantage of this type of connection is that if the joint surfaces are lightly greased, a vacuum seal is achieved. One of the drawbacks of using these joints is that contamination of the reacting system readily occurs if the solvents present in the reaction vessel dissolve the grease. In small-scale reactions this contamination can be particularly troublesome.

The small joints used in the microscale experimental organic laboratory, however, have the ease of assembly and physical integrity of research-grade, standard taper, ground-glass joints along with a number of important additional features. The joint dimensions are usually ⊤ 14/10. The conical vials in which most microscale reactions are carried out use this type of connecting system. Note that in addition to being ground to a standard taper on the inside surface of the throat of the vial, these vials also have a screw thread on the outside surface (Fig. 3.9).

This arrangement allows a standard taper male joint to be sealed to the reaction flask by a septum-type (open) plastic screw cap. The screw cap applies compression to a silicone rubber retaining O-ring positioned on the shoulder of the male joint (Fig. 3.10). The compression of the O-ring thereby achieves a greaseless gas-tight seal on the joint seam, while at the same time clamping the two pieces of equipment together. The ground joint provides both protection from intimate solvent contact with the O-ring and mechanical stability to the connection. The use of this type of connector leads to a further bonus during construction of an experimental setup. Because the individual sections are small, light, and firmly sealed together, the entire arrangement often can be mounted on the support rack by a single clamp. In conventional systems it is often necessary to use at least two clamps. This can easily lead to strain in the glass components unless considerable care is taken in the assembly process. Clamp strain is one of the major sources of experimental glassware breakage. The ability to single-clamp most microscale setups effectively eliminates this problem.

Note

When ground-glass joint surfaces are grease free it is important to disconnect joints soon after use (particularly with basic solutions) or they may become locked or "frozen" together.

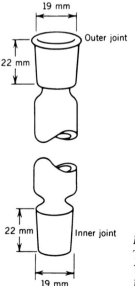

19 mm

Outer joint

22 mm

22 mm

Inner joint

19 mm

Figure 3.8 Standard taper joints (⊤). *(From Zubrick, James W. The Organic Chem Lab Survival Manual, 4th ed.; Wiley: New York, 1997. Reprinted by permission of John Wiley & Sons, Inc., New York.)*

Figure 3.9 Threaded female joint.

Joints of the size employed in these microscale experiments, however, are seldom a problem to separate if given proper care (*keep them clean!*).

Conical vials

Both the conical vials (3 and 5 mL) and the round-bottom flasks are designed to be connected via an O-ring compression cap installed on the male joint of the adjacent part of the system (see Fig. 3.1).

Condensers

Two types of condensers (air condensers and water-jacketed condensers) are available; in most cases the water-jacketed condenser can work well as an air condenser. Condensers are usually attached to 14/10 T-jointed reaction flasks. The upper female joints allow connection of the condenser to the 14/10 T drying tube and the 14/10 T capillary gas delivery tube (see Fig. 3.2).

Distillation heads

The simple Hickman still is used with an O-ring compression cap to carry out semi-micro simple or crude fractional distillations. The Hickman–Hinkle spinning band still uses a 3-cm fractionating column and routinely develops between five and six theoretical plates. The Hickman–Hinkle still is currently available with 14/10 T joints and can be conveniently operated with the 14/10 T 3- and 5-mL conical vials (see Figs. 3.1, 3.3, and 3.7). The still head is also available with an optional sidearm collection port.

Recrystallization tubes

Craig tubes are a particularly effective method for recrystallizing small quantities of reaction products. These tubes possess a nonuniform ground joint in the outer section. The substitution of Teflon for glass in the head makes these systems quite durable and much less susceptible to breakage during centrifugation (see Fig. 3.4).

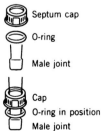

Septum cap

O-ring

Male joint

Cap
O-ring in position
Male joint   *Figure 3.10* Male joint with septum cap and O-ring.

Miscellaneous items

The Claisen head (see Fig. 3.3) is often used to facilitate the syringe addition of reagents to closed moisture-sensitive systems (such as Grignard reactions) via a septum seal in the vertical upper joint (see Fig. 3.18). This joint can also function to position the thermometer (using an adapter) in the well of a Hickman–Hinkle still (see Example [2A], p. 76). The Claisen adapter is also used to mount the drying tube in a protected position remote from the reaction chamber. The drying tube, in turn, is used to protect moisture-sensitive reaction components from atmospheric water vapor, while allowing a reacting system to remain unsealed. The capillary gas delivery tube is employed in transferring gases formed during reactions to storage containers (see Fig. 3.5 and 🌐 Chapter 3, Fig. 3.11W).

Gas chromatographic fraction collection items

For fraction collection the gas chromatographic (GC) collection tube is connected directly to the exit port of the GC detector through a stainless steel standard taper adapter. The collected sample is then transferred to a 0.1-mL conical vial for storage. The system is conveniently employed in the resolution and isolation of two-component mixtures (see Fig. 3.6).

STANDARD EXPERIMENTAL APPARATUS

Heating and stirring arrangements

It is important to be able to carry out microscale experiments at accurately determined temperatures. Very often, transformations are successful, in part, because of the ability to maintain precise temperature control. In addition, many reactions require reactants to be intimately mixed to obtain a substantial yield of product. Therefore, the majority of the reactions you perform in this laboratory will be conducted with rapid stirring of the reaction mixture.

Sand bath technique—hot plate calibration

A most convenient piece of equipment for heating or stirring or for performing both operations simultaneously on a microscale level is the hot-plate–magnetic stirrer. Heat transfer from the hot surface to the reaction flask is generally accomplished with a crystallizing dish containing a *shallow* layer of sand that can conform to the size and shape of the particular vessel employed. The temperature of the system is monitored by a thermometer embedded in the sand near the reaction vessel.

A successful procedure for determining the temperature inside the vial relative to the bath temperature is to mount a second thermometer in a vial containing 2 mL of high-boiling silicone oil. The vial temperature is then measured at various sandbath temperatures and the values are entered on graphs of vial temperatures versus hot-plate settings and bath temperatures versus hot-plate settings (see Fig. 3.11 and 🌀 Chapter 3, Fig. 3.5W) for your particular hot-plate system (see also section on Metal Heat-Transfer Devices, p. 25). These data will save considerable time when you bring a reaction system to operating temperature. When you first enter the laboratory, it is advisable to adjust the temperature setting on the hot-plate stirrer with the heating device, or bath, in place. The setting is determined from your control setting–temperature calibration curve. This procedure will allow the heated bath to reach a relatively constant temperature by the time it is required. You will then be able to make small final adjustments more quickly, if necessary.

Note

> Heavy layers of sand act as an insulator on the hot-plate surface, which can damage the heating element at high temperature settings. When temperatures over 150 °C are required, it is especially important to use the minimum amount of sand.

Recording the weight of sand used and the size of the crystallizing dish will help to make the graph values more reproducible.

The high sides of the crystallizing dish protect the apparatus from air drafts, and so the dish also operates somewhat as a hot-air bath. Heating can be made even more uniform by covering the crystallizing dish with aluminum foil (see Fig. 3.12 and 🌀 Chapter 3, Fig. 3.1W). This procedure works well, but is a bit awkward and is required in only a few instances.

The insulating properties of sand provide a readily available variable heat source because the temperature of the sand is higher deeper in the bath; thus, the depth of sand used in the bath is exceedingly important. **The depth should always be kept to a minimum, in the range of 10–15 mm.** Finally, sand baths offer a significant safety advantage over oil baths. Individual grains of sand are so small that they have little heat capacity and thus are less likely to burn the chemist in the event of a spill.

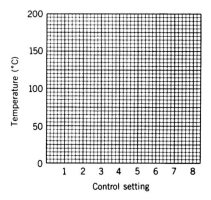

Figure 3.11 Plot your bath and/or vial temperature (°C) versus hot-plate control setting.

Metal heat-transfer devices

An alternative to the sand bath is a heat-transfer system that employs copper tube plates or aluminum metal blocks drilled to accommodate the different reaction vials and flasks (🌀 Chapter 3, Fig. 3.3W).

Stirring

Stirring the reaction mixture in a conical vial is carried out with Teflon-coated magnetic spin vanes, and in round-bottom flasks with Teflon-coated magnetic stirring bars (see Fig. 3.12 and 🌀 Chapter 3, Fig. 3.1W). It is important to put the reaction flask as close to the center and to the bottom surface of the crystallizing dish as possible when using magnetic stirring. This arrangement is a good practice in general, as it leads to using the minimum amount of sand needed in a sand bath.

If the reaction does not require elevated temperatures, but needs only to be stirred, the system can be assembled without the heat-transfer device (sand bath or metal plate). Some stirred reactions, on the other hand, require cooling. In these cases a crystallizing dish filled with ice water, or with ice water and salt, if lower temperatures are called for, will provide the correct environment.

Reflux apparatus

To bring about a successful reaction between two substances, it is often necessary to mix the materials together intimately and to maintain a specific temperature. The mixing operation is conveniently achieved by dissolving the materials in a solvent in which they are mutually soluble. If the reaction is carried

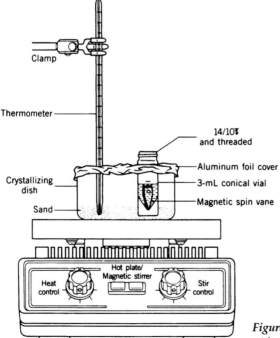

Figure 3.12 Hot-plate–magnetic stirrer with sand bath and reaction vial.

out in solution under reflux conditions, the choice of solvent can be used to control the temperature of the reaction. Many organic reactions involve the use of a reflux apparatus in one arrangement or another.

What do we mean by *reflux?* The term means to "return," or "run back." This return is exactly how the reflux apparatus functions. When the temperature of the reaction system is raised to the solvent's boiling point (constant temperature), all vapors are condensed and returned to the reaction flask or vial; this operation is not a distillation and the liquid phase remains at a stable maximum temperature. In microscale reactions, two basic types of reflux condensers are utilized: the air-cooled condenser, or air condenser (🐚 Chapter 3, Fig. 3.6W), and the water-jacketed condenser (see Fig. 3.13 and 🐚 Chapter 3, Fig. 3.7W). The air condenser condenses solvent vapors on the cool vertical wall of an extended glass tube that dissipates the heat by contact with the laboratory room air. This arrangement functions quite effectively with liquids boiling above 150 °C. Indeed, a simple test tube can act as a reaction chamber and air condenser all in one unit, and many simple reactions can be most easily carried out in test tubes.

Air condensers can occasionally be used with lower boiling systems; however, the water-jacketed condenser is more often employed in these situations. The water-jacketed condenser employs flowing cold water to remove heat from the vertical column and thus facilitate vapor condensation. It is highly effective at condensing vapor from low-boiling liquids.

Both styles of condensers accommodate various sizes of reaction flasks and are available with 14/10ᵀ standard taper joints. The tops of both condenser columns have a female 14/10ᵀ joint.

In refluxing systems that do not require significant mixing or agitation, the stirrer (magnetic spin vane or bar) usually is replaced by a "boiling stone."

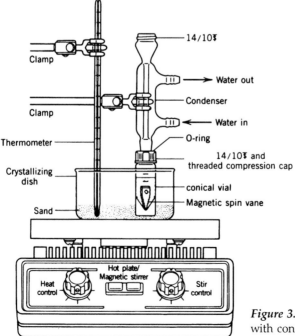

Figure 3.13 Water-jacketed condenser with conical vial, arranged for heating and magnetic stirring.

These sharp-edged stones possess highly fractured surfaces that are very efficient at initiating bubble formation as the reacting medium approaches the boiling point. The boiling stone acts to protect the system from disastrous boilovers and also reduces "bumping." (Boiling stones should be used only once and must **never** be added to a hot solution. In the first case, the vapor cavities become filled with liquid upon cooling, and thus a boiling stone becomes less effective after its first use. In the second case, **adding the boiling stone to the hot solution may suddenly start an uncontrollable boilover**).

Distillation apparatus

Distillation is a laboratory operation used to separate substances that have different boiling points. The mixture is heated, vaporized, and then condensed; the early fractions of condensate are enriched in the more volatile component. Unlike the reflux operation, in distillations none, or only a portion, of the condensate is returned to the flask where vaporization is taking place. Many distillation apparatus have been designed to carry out this basic operation. They differ mainly in small features that are used to solve particular types of separation problems. In carrying out these distillations the choice of still depends to a large degree on the difficulty of the separation required (generally, how close are the boiling points in the mixture to be separated?).

The Hickman still head (Fig. 3.14) is ideally suited for simple distillations. This system has a 14/10 male joint for connection to conical vials or round-

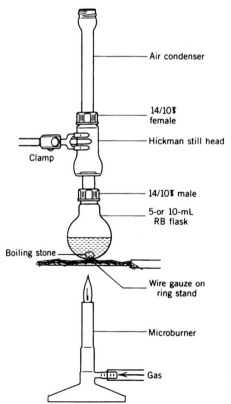

- Air condenser
- 14/10 female
- Hickman still head
- Clamp
- 14/10 male
- 5- or 10-mL RB flask
- Boiling stone
- Wire gauze on ring stand
- Microburner
- Gas

Figure 3.14 Hickman still head and air condenser with 5-mL round-bottom flask, arranged for microburner heating.

bottom flasks. The still head functions as both an air condenser and a condensate trap. For a detailed discussion of this piece of equipment see Distillation Example [2A], p. 76. The simple Hickman still has been modified (see Fig. 3.15) with a spinning band. The still continues to function in much the same way as the simple Hickman still, but a tiny Teflon spinning band is now mounted in a slightly extended section between the male joint and the collection collar. When the band is spun at 1500 rpm by a magnetic-stirring hot plate, this still functions as an effective short-path fractional distillation column (see Distillation Example [3B], p. 87. In addition, this modified system has a built-in thermometer well that allows reasonably accurate measurement of vapor temperatures plus a sidearm port for removing distillate.

The most powerful microscale distillation system currently available is the 2.5-in. vacuum-jacketed microscale spinning-band distillation column (see Fig. 3.16 and Distillation Example [3A], p. 83 for description and details). This still is designed for conventional downward distillate collection and nonstopcock reflux control. The column is rated at ~10 theoretical plates.

For a discussion of reduced pressure distillation see 🌑 Distillation.

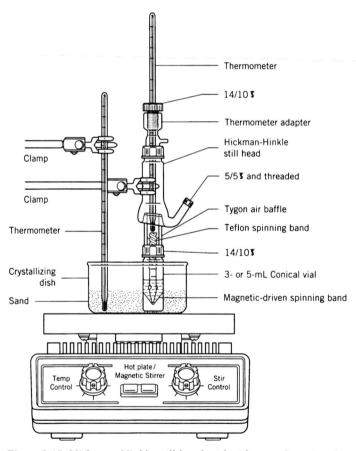

Figure 3.15 Hickman–Hinkle still head with side-port 3- or 5-mL conical vial, Teflon spinning band, and thermometer adapter and arranged for heating and magnetic stirring.

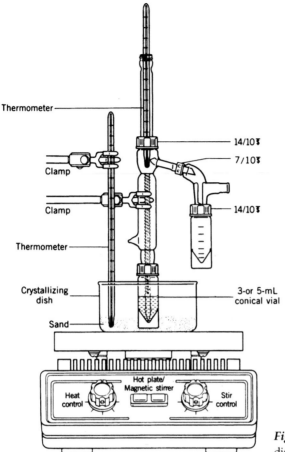

Thermometer

14/10⫪

7/10⫪

Clamp

Clamp

14/10⫪

Thermometer

Crystallizing
dish

3-or 5-mL
conical vial

Sand

Hot plate/
Magnetic stirrer

Heat
control

Stir
control

Figure 3.16 Micro spinning band
distillation column (2.5 in).

Moisture-protected reaction apparatus

Many organic reagents react rapidly and preferentially with water. *The success or failure of many experiments depends to a large degree on how well atmospheric moisture is excluded from the reaction system.* The "drying tube," which is packed with a desiccant such as anhydrous calcium chloride, is a handy way to carry out a reaction in an apparatus that is not totally closed to the atmosphere, but that is reasonably well protected from water vapor. The microscale apparatus described here are designed to be used with the 14/10⫪ drying tube. The reflux condensers discussed earlier are constructed with female 14/10⫪ joints at the top of the column, which allows convenient connection of the drying tube if the refluxing system is moisture sensitive (see Fig. 3.17).

Because many reactions are highly sensitive to moisture, successful operation at the microscale level can be rather challenging. If anhydrous reagents are to be added after an apparatus has been dried and assembled, it is important to be able to introduce these reagents without exposing the system to the atmosphere, particularly when operating in a humid atmosphere. In room-temperature reactions that do not need refluxing, adding anhydrous reagents

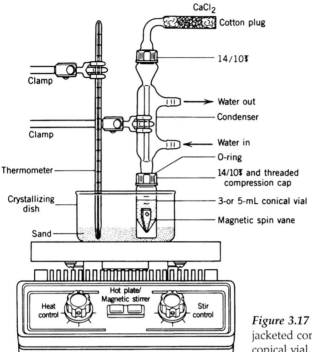

Figure 3.17 Moisture-protected water-jacketed condenser with 3- or 5-mL conical vial, arranged for heating and magnetic stirring.

is best accomplished by use of the microscale Claisen head adapter. The adapter has a vertical screw-threaded standard taper joint that will accept a septum cap. The septum seal allows syringe addition of reagents and avoids the necessity of opening the apparatus to the laboratory atmosphere (see Fig. 3.18).

Specialized pieces of equipment

COLLECTION OF GASEOUS PRODUCTS

Some experiments lead to gaseous products. The collection, or trapping, of gases is conveniently carried out by using the capillary gas delivery tube. This item is designed to be attached directly to a 1- or 3-mL conical vial (see Fig. 3.19), or to the female 14/10₮ joint of a condenser connected to a reaction flask or vial (🔖 Chapter 3, Fig. 3.11W). The tube leads to the collection system, which may be a simple, inverted, graduated cylinder; a blank-threaded septum joint; or an air condenser filled with water (if the gaseous products are not water-soluble). The 0.1-mm capillary bore considerably reduces dead volume and increases the efficiency of product transfer.

COLLECTION OF GAS CHROMATOGRAPHIC EFFLUENTS

The trapping and collection of gas chromatographic liquid fractions become particularly important exercises in microscale experiments. When yields

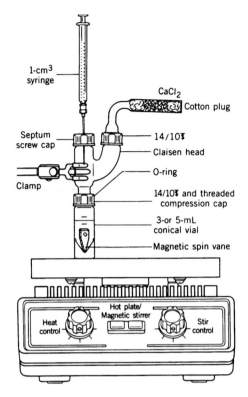

Figure 3.18 Moisture-protected Claisen head with 3- or 5-mL conical vial, arranged for syringe addition and magnetic stirring.

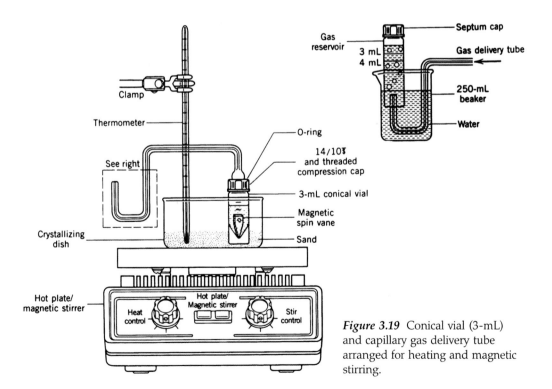

Figure 3.19 Conical vial (3-mL) and capillary gas delivery tube arranged for heating and magnetic stirring.

of a liquid product are less than 100 μL, conventional distillation, even using microscale equipment, is impractical. In this case, preparative gas chromatography replaces conventional distillation as the route of choice to product purification. The ease and efficiency of carrying out this operation is greatly facilitated by employing the 5/5ᵀ collection tube and the 0.1-mL 5/5ᵀ conical collection vial (See Fig. 3.6, p. 20) (🌐 Chapter 3, Fig. 3.12W).

MICROSCALE LAWS

Rules of the trade for handling organic materials at the microscale level

Now that we have briefly looked at the equipment we will be using to carry out microscale organic reactions, let us examine the specific techniques that are used to deal with the small quantities of material involved. Microscale synthetic organic reactions, as defined by Cheronis,[1] start with 15–150 mg of the limiting reagent. These quantities sound small, and they are. Although 150 mg of a light, powdery material will fill half a 1-mL conical vial, you will have a hard time observing 15 mg of a clear liquid in the same container, even with magnification. This volume of liquid, on the other hand, is reasonably easy to observe if it is in a 0.1-mL conical vial. A vital part of the game of working with small amounts of materials is to become familiar with microscale techniques and to practice them as much as possible in the laboratory.

Rules for working with liquids at the microscale level

1. *Liquids are never poured at the microscale level.* Liquid substances are transferred by pipet or syringe. As we are working with small, easy-to-hold glassware, the best way to transfer liquids is to hold both containers with the fingers of one hand, with the mouths as close together as possible. The free hand is then used to operate the pipet (syringe) to withdraw the liquid and make the transfer. This approach reduces to a minimum the time that the open tip is not in, or over, one vessel or the other. We use three different pipets and two standard syringes to perform most experiments involving liquids. **This equipment can be a prime source of contamination.** Be very careful to thoroughly clean the pipets and syringes after each use.

a. *Pasteur pipet (often called a capillary pipet).* A simple glass tube with the end drawn to a fine capillary. These pipets can hold several milliliters of liquid (Fig. 3.20*a*) and are filled using a small rubber bulb or one of the very handy, commercially available pipet pumps. Because many transfers are made with Pasteur pipets, it is suggested that several of them be calibrated for ap-

---

[1]Cheronis, N. D., *Semimicro Experimental Organic Chemistry*; Hadrion Press: New York, 1958.

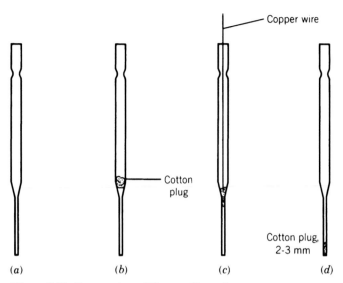

Figure 3.20 Preparation of Pasteur filter pipet.

proximate delivery of 0.5, 1.0, 1.5, and 2.0 mL of liquid. This calibration is easily done by drawing the measured amount of a liquid from a graduated cylinder and marking the level of the liquid in the pipet. This mark can be made with transparent tape, or by scratching with a file. Indicate the level with a marking pen before trying to tape or file the pipet.

b. *Pasteur filter pipet.* A very handy adaptation of the Pasteur pipet is a filter pipet. This pipet is constructed by taking a small cotton ball and placing it in the large open end of the standard Pasteur pipet. Hold the pipet vertically and tap it gently to position the cotton ball in the drawn section of the tube (Fig. 3.20b). Now form a plug in the capillary section by pushing the cotton ball down the pipet with a piece of copper wire (Fig. 3.20c). Finish by seating the plug flush with the end of the capillary (Fig. 3.20d). The optimum-size plug will allow easy movement along the capillary while it is being positioned by the copper wire. Compression of the cotton will build enough pressure against the walls of the capillary (once the plug is in position) to prevent the plug from slipping while the pipet is filled with liquid. If the ball is too big, it will wedge in the capillary before the end is reached, and wall pressure will be so great that liquid flow will be shut off. Even some plugs that are loose enough to be positioned at the end of the capillary will still have developed sufficient lateral pressure to make the filling rate unacceptably slow. If the cotton filter, however, is positioned too loosely, it may be easily dislodged from the pipet by the solvent flow. These plugs can be quickly and easily inserted with a little practice. Once in place, the plug is rinsed with 1 mL of methanol and 1 mL of hexane, and dried before use.

There are two reasons for placing the cotton plug in the pipet. First, it solves a particular problem with the transfer of volatile liquids via the standard Pasteur pipet: the rapid buildup of back pressure from solvent vapors in the rubber bulb. This pressure quickly tends to force the liquid back out of the pipet and can cause valuable product to drip on the bench top. The cotton plug tends to resist this back pressure and allows much easier control of the solution once it is in the pipet. The time-delay factor becomes particularly important when the Pasteur filter pipet is employed as a microseparatory funnel (see the discussion on extraction techniques in Technique 4, p. 90).

Second, each time a transfer of material is made, the material is automatically filtered. This process effectively removes dust and lint, which are constant problems when working at the microscale level. A second stage of filtration may be obtained by employing a disposable filter tip on the original Pasteur filter pipet as described by Rothchild.[2]

c. *Automatic pipet (considered the Mercedes–Benz of pipets).* Automatic pipets quickly, safely, and reproducibly measure and dispense specific volumes of liquids. These pipets are particularly valuable at the microscale level, because they generate the precise and accurate liquid measurements that are absolutely necessary when handlisng only microliters of a liquid. The automatic pipet adds considerable insurance for the success of an experiment, since any liquid can be efficiently measured, transferred, and delivered to the reaction flask.

The automatic pipet consists of a calibrated piston pipet with a specially designed disposable plastic tip. It is possible to encounter any one of three pipet styles: single volume, multirange, or continuously adjustable (see Fig. 3.21). The first type is calibrated to deliver only a single volume. The second type is adjustable to two or three predetermined delivery volumes. The third type is the most versatile; it can be set by the user to deliver any volume within the range of the pipet. Obviously, the price of these valuable laboratory tools goes up with increasing features. Automatic pipets are expensive, and usually must be shared in the laboratory. Treat them with respect!

The automatic pipet is designed so that the liquid comes in contact only with the disposable tip.

- Never load the pipet without the tip in place.
- Never immerse the tip completely in the liquid that is being pipetted.
- Always keep the pipet vertical when the tip is attached.
- If an air bubble forms in the tip during uptake, return the liquid, discard the tip, and repeat the sampling process.

If these three rules are followed, most automatic pipets will give many years of reliable service. A few general rules for improving reproducibility with an automatic pipet should also be followed:

- Try to use the same uptake and delivery motion for all samples. Smooth depression and release of the piston will give the most consistent results. Never allow the piston to snap back.
- *Always* depress the piston to the first stop before inserting the tip into the liquid. If the piston is depressed after submersion, formation of an air bubble in the tip becomes likely, which will result in a filling error.

---

[2]Rothchild, R. J. *Chem. Educ.* **1990**, *67*, 425.

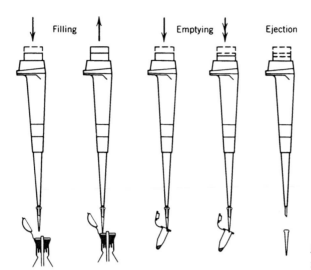

Figure 3.21 Operation of automatic delivery pipet.

- *Never* insert the tip more than 5 mm into the liquid. It is good practice not to allow the body of the pipet to contact any surface, or bottle neck, that might be wet with a chemical.
  - If an air bubble forms in the tip during uptake, return the fluid, discard the tip, and repeat the sampling process.

d. *Syringes.* Syringes are particularly helpful for transferring liquid reagents or solutions to sealed reaction systems from sealed reagent or solvent reservoirs. Syringe needles can be inserted through a septum, which avoids opening the apparatus to the atmosphere. Syringes are also routinely employed in the determination of ultramicro boiling points (10-μL GC syringe). It is critically important to clean the syringe needle after each use. Effective cleaning of a syringe requires as many as a dozen flushes. For many transfers, the microscale laboratory uses a low-cost glass 1-mL insulin syringe in which the rubber plunger seal is replaced with a Teflon seal (ACE Glass). For preparative GC injections, the standard 50- or 100-μL syringes are preferred (see Technique 1).

2. *Liquid volumes may be converted easily to mass measures by the following relationship:*

$$\text{Volume (mL)} = \frac{\text{mass (g)}}{\text{density (g/mL)}}$$

3. *Work with liquids in conical vials,* and work in a vial whose capacity is approximately twice the volume of the material it needs to hold. The trick here is to reduce the surface area of the flask in contact with the sample to an absolute minimum. A conical vial is thus better than the spherical surface of the conventional round-bottom flask.

Liquids may also be weighed directly. A tared container (vial) should be used. After addition of the liquid, the vial should be kept capped throughout the weighing operation. This procedure prevents loss of the liquid by evaporation. If the density of the liquid is known, the approximate volume of the liquid should be transferred to the container using an automatic delivery pipet or a calibrated Pasteur pipet. Use the above expression relating density, mass, and volume to calculate the volume required by the measured mass. Adjustment of the mass to give the desired value can then be made by adding or removing small amounts of liquid from the container by Pasteur pipet.

Note

> Before you leave the balance area, be sure to replace all caps on reagent bottles and clean up any spills. A balance is a precision instrument that can easily be damaged by contamination.

Rules for working with solids at the microscale level

1. *General considerations.* Working with a crystalline solid is much easier than working with the equivalent quantity of a liquid. Unless the solid is in solution, a spill on a clean glass working surface usually can be recovered quickly and efficiently. Be careful, however, when working with a solution. Treat solutions as you would a pure liquid.

2. *Transfer of solids.* Solids are normally transferred with a microspatula, a technique that is not difficult to develop.

3. *Weighing solids at the milligram level.* Electronic balances can automatically tare an empty vial. Once the vial is tared, the reagent is added in small portions. The weight of each addition is instantly registered; material is added until the desired quantity has been transferred.

Solids are best weighed in glass containers (vials or beakers), in plastic or aluminum weighing trays ("boats"), or on glazed weighing paper. Filter paper or other absorbent materials are not good choices: small quantities of the weighed material will often stick to the fibers of the paper, and vice versa.

THE LABORATORY NOTEBOOK

Writing is the most important method chemists use to communicate their work. It begins with the record kept in a laboratory notebook. An experiment originally recorded in the laboratory notebook can become the source of information used to prepare scientific papers published in journals or presented at meetings. For the industrial chemist, these written records are critical in obtaining patent coverage for new discoveries.

It is important that you learn to keep a detailed account of your work. A laboratory notebook has several key components. Note how each component is incorporated into the example that follows.

KEY COMPONENTS OF A LABORATORY EXPERIMENT WRITE-UP

1. Date experiment was conducted
2. Title of experiment
3. Purpose for running the reaction
4. Reaction scheme
5. Table of reagents and products
6. Details of procedure used
7. Characteristics of the product(s)
8. References to product or procedure (if any)
9. Analytical and spectral data
10. Signature of person performing the experiment and that of a witness, if required

In reference to point 6, it is the obligation of the person doing the work to list the equipment, the amounts of reagents, the experimental conditions, and the method used to isolate the product. Any color or temperature changes should be carefully noted and recorded.

Several additional points can be made about the proper maintenance of a laboratory record.

11. A hardbound, permanent notebook is essential.
12. Each page of the notebook should be numbered in consecutive order. For convenience, an index at the beginning or end of the book is recommended and pages should be left blank for this purpose.
13. If a page is not completely filled, an "X" should be used to show that no further entry was made.
14. Data are always recorded directly into the notebook, *never* on scrap paper! Always record your data in ink. If a mistake is made, draw a neat line through the word or words so that they remain legible. Do not completely obliterate anything; you might learn from your mistakes, if you can read them later.
15. Make the record clear and unambiguous. Pay attention to grammar and spelling.
16. In industrial research laboratories, your signature, as well as that of a witness, is required, because the notebook may be used as a legal document.
17. Always write and organize your work so that someone else could come into the laboratory and repeat your directions without confusion or uncertainty. *Completeness* and *legibility* are key factors.

Most of you are newcomers to the organic laboratory, and the reactions you will be performing have probably been worked out and checked in detail. Because of this, your instructor may not require you to keep your notebook in such a meticulous fashion. For example, when you describe the procedure (item 6), it may be acceptable to make a clear reference to the material in the laboratory manual and to note any modifications or deviations from the prescribed procedure. In some cases, it may be more practical to use an outline method. In any event, the following example should be studied carefully. It may be used as a reference when detailed records are important in your work. It is more important to record what you observed and what you actually did, than to record what you were supposed to observe and what you were supposed to do.

Note

Because of its length, the example here is typed. Notebooks are usually handwritten. Many chemists, however, now use computers to record their data.

The circled numbers refer to the list on page 37.

EXAMPLE OF A LABORATORY NOTEBOOK ENTRY

16 Aug. 1985 ①
PREPARATION OF DIPHENYL SUCCINATE ②

$$\begin{matrix} CH_2CO_2H \\ | \\ CH_2CO_2H \end{matrix} + 2\ C_6H_5OH + POCl_3 \rightarrow \begin{matrix} CH_2CO_2C_6H_5 \\ | \\ CH_2CO_2C_6H_5 \end{matrix} + HPO_3 + 3\ HCl \left.\right\} ④$$

③ —— Diphenyl succinate is being prepared as one of a series of dicarboxylic acid esters that are to be investigated as growth stimulants for selected fungi species.

⑧ —— This procedure was adapted from that reported by Daub, G. H., and Johnson, *W. S. Organic Syntheses*, Wiley: New York, 1963; Collect. Vol. IV, p. 390.

Physical Properties of Reactants and Products

⑤ ——

COMPOUND	MW[a]	AMOUNTS	mmol	mp (C°)	bp (°C)
Succinic acid	118.09	118 mg	1.0	182	
Phenol	94.4	188 mg	2.0	40–42	182
Phosphorous oxychloride	153.33	84 μL	0.9		105.8
Diphenyl succinate	270.29			121	

[a]MW = molecular weight.

In a 3.0-mL conical vial containing a magnetic spin vane and equipped with a reflux condenser protected by a calcium chloride drying tube were ⑥— placed succinic acid (118 mg, 1.0 mmol), phenol (188 mg, 2.0 mmol), and phosphorous oxychloride (84 μL 0.9 mmol). The reaction mixture was heated with stirring at 115 °C in a sand bath in the **hood** for 1.25 h. It was necessary to conduct the reaction in the **hood,** because hydrogen chloride (HCl) gas evolved during the course of the reaction. The drying tube was removed, toluene (0.5 mL) was added through the top of the condenser using a Pasteur pipet, and the drying tube was replaced. The mixture was then heated for an additional 1 hour at 115 °C.

The hot toluene solution was separated from the red syrupy residue of phosphoric acid using a Pasteur pipet. The toluene extract was filtered by gravity using a fast-grade filter paper and the filtrate was collected in a 10-mL Erlenmeyer flask. The phosphoric acid residue was then extracted with two additional 1.0-mL portions of hot toluene. These extracts were also separated using the Pasteur pipet and filtered, and the filtrate was collected in the same Erlenmeyer flask. The combined toluene solutions were concentrated to a volume of approximately 0.6 mL by warning them in a sand bath under a gentle stream of nitrogen (N_2) gas in the **hood.** The pale yellow liquid residue was then allowed to cool to room temperature; the diphenyl succinate precipitated as colorless crystals. The solid was collected by vacuum filtration using a Hirsch funnel, and the filter cake was washed with three 0.5-mL portions of cold diethyl ether. The product was dried in a vacuum oven at 30 °C (3 mm Hg) for 30 min.

The 181 mg (67%) of diphenyl succinate had an mp of 120–121 °C (lit. value 121 °C: *CRC Handbook of Chemistry and Physics*, 60th ed.; CRC Press: Boca Raton, FL, 1979; #S 197, p C-501).

The IR spectrum exhibits the expected peaks for the compound. [*At this point, the data may be listed, or the spectrum attached to a separate page of the notebook.*]

Marilyn C. Waris

witnessed by
D. Jeanne d'Arc Mailhiot *16/Aug./1985*

CALCULATING YIELDS

Almost without exception, in each of the experiments presented in this text, you are asked to calculate the percentage yield. For any reaction, it is always important for the chemist to know how much of a product is actually produced (experimental) compared to the theoretical (maximum) amount that could have been formed. The percentage yield is calculated on the basis of the relationship.

$$\% \text{ yield} = \frac{\text{actual yield (experimental)}}{\text{theoretical yield (calculated maximum)}} \times 100$$

The percentage yield is generally calculated on a weight (gram or milligram) or on a mole basis. In the present text, the calculations are made using milligrams.

Several steps are involved in calculating the percentage yield.

Step 1. Write a *balanced* equation for the reaction. For example, consider the Williamson synthesis of propyl *p*-tolyl ether.

$$CH_3-\langle\bigcirc\rangle-OH + CH_3CH_2CH_2-I \xrightarrow[\text{(C}_4\text{H}_9)_4\text{N}^+,\text{ Br}^-]{\text{NaOH}} CH_3-\langle\bigcirc\rangle-O-(CH_2)_2CH_3 + Na^+, I^-$$

p-Cresol   Propyl iodide   Propyl *p*-tolyl ether

Physical Properties of Reactants

COMPOUND	MW	AMOUNTS	mmol	d
p-Cresol	108.15	160 μL	1.56	1.5312
25% (by weight) NaOH soln	40.0	260 μL	~1.6	
Tetrabutylammonium bromide	322.28	18 mg	0.056	
Propyl iodide	169.99	150 μL	1.54	1.5058

Step 2. Identify the *limiting* reactant. The ratio of reactants is calculated on a millimole (or mole) basis. In the example, 1.56 mmol of *p*-cresol and ca. 1.6 mmol of sodium hydroxide are used, compared to 1.54 mmol of propyl iodide, which is therefore the limiting reagent. The tetrabutylammonium bromide is not considered because it is used as a catalyst—it is neither incorporated into the product nor consumed in the reaction. The calculation of the theoretical yield is thus based on the amount of propyl iodide, 1.54 mmol.

Step 3. Calculate the *theoretical* (maximum) amount of the product that could be obtained for the conversion, based on the limiting reactant. Here, one mole of propyl iodide produces one mole of the propyl *p*-tolyl ether. Therefore, the maximum amount of propyl *p*-tolyl ether (molecular weight = 150.2) that can be produced from 1.54 mmol of propyl iodide is 1.54 mmol, or 231 mg.

Step 4. Determine the *actual* (experimental) yield (milligrams) of product isolated in the reaction. This amount is invariably less than the theoretical quantity, unless the material is impure (one common contaminant is water). For example, student yields for the preparation of propyl *p*-tolyl ether average 140 mg.

Step 5. Calculate the *percentage yield* using the weights determined in steps 3 and 4. The percentage yield is then

$$\% \text{ yield} = \frac{140 \text{ mg (actual)}}{231 \text{ mg (theoretical)}} \times 100 = 60.6\%$$

4

Determination of Physical Properties

Determination of physical properties is important for substance identification and as an indication of material purity. Historically, the physical constants of prime interest have included boiling point, density, and refractive index in liquids and the melting point in solids. In special cases, optical rotation and molecular weight determinations may be required. Today, with the widespread availability of spectroscopic instrumentation, powerful new techniques may be applied to the direct identification and characterization of materials, including the analysis of individual components of very small quantities of complex mixtures. The sequential measurement of the infrared (IR) and mass spectrometric (MS) characteristics of a substance resolved "on the fly" by capillary gas chromatography (GC) can be quickly determined and interpreted. This particular combination (GC-IR-MS), which stands out among a number of hyphenated techniques becoming available, is perhaps the most powerful system yet developed for molecular identification. The rapid development of high-field multinuclear magnetic resonance (NMR) spectrometers has added another powerful dimension to identification techniques. NMR sensitivity, however, is still considerably lower than that of either IR or MS. The IR spectrum alone, obtained with one data point per wavenumber can add more than 4000 measurements to the few classically determined properties. *Indeed, even compared to high-resolution MS and pulsed ^1H and ^{13}C NMR, the infrared spectrum of a material remains a powerful set of physical properties (transmission elements) available to the organic chemist for the identification of an unknown compound.*[1]

Simple physical constants are determined mainly to assist in establishing the purity of known materials. Because the boiling point or the melting point of a material can be very sensitive to small quantities of impurities, these data can be particularly helpful in determining whether a starting material needs further purification or whether a product has been isolated in acceptable pu-

---

[1]Griffiths, P.R.; de Haseth, J. *Fourier Transform Infrared Spectrometry;* Wiley: New York, 1986.

rity. Gas (GC), high-performance liquid (HPLC), and thin-layer (TLC) chromatography, however, now provide more powerful purity information when such data are required. When a new composition of matter has been formed, an elemental (combustion) analysis is normally reported if sufficient material is available for this destructive analysis. For new substances we are, of course, interested in establishing not only the identity, but also the molecular structure of the materials. In this situation other modern techniques (such as ^{1}H and ^{13}C NMR spectroscopy, high-resolution MS, and single-crystal X-ray diffraction) can provide powerful structural information.

When comparisons are made between experimental data and values obtained from the literature, it is essential that the latter information be obtained from the most reliable sources available. Certainly, judgment, which improves with experience, must be exercised in accepting any value as a standard. The known classical properties of a large number of compounds are found in the *CRC Handbook of Chemistry and Physics* and the *Merck Index*. The *Aldrich Catalog Handbook of Fine Chemicals* is also a readily available, inexpensive source. These reference works list physical properties for inorganic, organic, and organometallic compounds. The *Aldrich Catalog* also references IR and NMR data for a large number of substances. New editions of the *CRC Handbook* and the *Aldrich Catalog* are published each year.

I. *Liquids*

ULTRAMICRO BOILING POINT

Upon heating, the vapor pressure of a liquid increases, though in a nonlinear fashion. When the pressure reaches the point where it matches the local pressure, the liquid boils. That is, it spontaneously begins to form vapor bubbles, which rapidly rise to the surface. If heating is continued, both the vapor pressure and the temperature of the liquid will remain constant until the substance has been completely vaporized (Fig. 4.1).

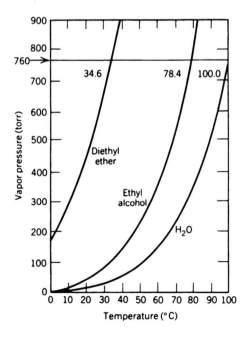

Figure 4.1 Vapor pressure curves. (*From Brady, J. E.; Humiston, G. E. General Chemistry, 3d ed; Wiley: New York, 1982. (Reprinted by permission of John Wiley & Sons, New York.)*

Because microscale preparations generally yield about 30–70 μL of liquid products, using only 5 μL or less of material for boiling point measurements is highly desirable. The modification of the earlier Wiegand ultramicro boiling-point procedure[2] to the ultramicro procedure described here has established that reproducible and reasonably accurate (±1 °C) boiling points can be observed on 3–4 μL of most liquids thermally stable at the required temperatures.

Procedure

Ultramicro boiling points can be conveniently determined in standard (90-mm-length) Pyrex glass capillary melting-point tubes. The melting-point tube replaces the conventional 3- to 4-mm (o.d.) tubing used in the Siwoloboff procedure.[3] The sample (3–4 μL) is loaded into the melting-point capillary via a 10-μL syringe and centrifuged to the bottom if necessary. A small glass bell replaces the conventional melting-point tube as the bubble generator in micro boiling-point determinations. It is formed by heating 3-mm (o.d.) Pyrex tubing with a microburner and drawing it out to a diameter small enough to be readily fit inside the melting-point capillary. A section of the drawn capillary is fused and then cut to yield two small glass bells approximately 5 mm long (Fig. 4.2a). It is important that the fused section be reasonably large, because it is more than just a seal. The fused glass must add enough weight to the bell that it will firmly seat itself in the bottom of the melting-point tube.

An alternative technique for preparing the glass bells follows: heat the midsection of an open-ended melting point capillary tube and then draw the glass to form a smaller capillary section. This section is then broken approximately in the middle and each open end is sealed. The appropriate length for the

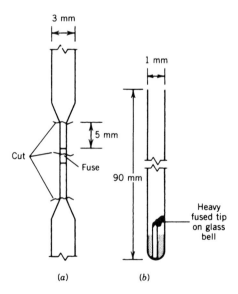

Figure 4.2   (a) Preparation of small glass bell for ultramicro boiling-point determination. (b) Ultramicro boiling-point assembly. *(From Mayo, D. W.; Pike, R. M.; Butcher, S. S.; Meredith, M. L. J. Chem. Educ. 1985, 62, 1114.)*

[2]Wiegand, C. *Angew. Chem.* **1955**, *67*, 77. Mayo, D. W.; Pike, R. M.; Butcher, S. S.; Meredith, M. L. *J. Chem. Educ.* **1985**, *62*, 1114.
[3]Siwoloboff, A. *Berichte* **1886**, *19*, 795.

bell is then broken off. Thus, two bells are obtained, one from each section. The sealing process (be sure that a significant section of glass is fused during the tube closure to give the bell enough weight) can be repeated on each remaining glass section and thus a series of bells can be prepared in a relatively short period.

A glass bell is now inserted into the loaded melting-point capillary, open end first (down), and allowed to fall (centrifuged if necessary) to the bottom. The assembled system (Fig. 4.2*b*) is then inserted onto the stage of a Thomas-Hoover Uni-Melt Capillary Melting Point Apparatus (Fig. 4.3)[4] or similar system (such as a Mel-Temp).

The temperature is rapidly raised to 15–20 °C below the expected boiling point (the temperature should be monitored carefully in the case of unknown substances), and then adjusted to a maximum rise rate of 2 °C/min and heated until a *fine stream* of bubbles is emitted from the glass bell. The heat control is

Figure 4.3 Thomas-Hoover melting-point determination device. (*Courtesy of Thomas Scientific, Swedesboro, NJ.*)

[4]Thomas Scientific, 989 High Hill Road, P.O. Box 99, Swedesboro, NJ 08085.

then adjusted to drop the temperature. The boiling point is recorded at the point where the last escaping bubble collapses (i.e., when the vapor pressure of the substance equals the atmospheric pressure). The heater is then rapidly adjusted to again raise the temperature at 2 °C/min and induce a second stream of bubbles. This procedure may then be repeated several times. *A precise and sensitive temperature control system is essential to the successful application of this cycling technique, but it is not essential for obtaining satisfactory boiling-point data.*

Utilization of the conventional melting-point capillary as the "boiler" tube has the particular advantage that the boiling point of a liquid can readily be determined using a conventional melting-point apparatus. The illumination and magnification available make the observation of rate changes in the bubble stream easily seen. Inexpensive 10-µL GC injection syringes appear to be the most successful instrument to use for transferring the small quantities of liquids involved. The 3-in. needles normally supplied with the 10-µL barrels will not reach the bottom of the capillary; liquid samples deposited on the walls of the tube, however, are easily and efficiently moved to the bottom by centrifugation. After the sample is packed in the bottom of the capillary tube, the glass bell is introduced. The glass bell is necessary because a conventional Siwoloboff fused-capillary insert would extend beyond the top of the melting-point tube; thus, capillary action between the "boiler" tube wall and the capillary insert would draw most of the sample from the bottom of the tube up onto the walls. This effect often precludes the formation of the requisite bubble stream.

Little loss of low-boiling liquids occurs (see Table 4.1). Furthermore, if the boiling point is overrun and the sample is suddenly evaporated from the bottom of the "boiler" capillary, it will rapidly condense on the upper (cooler) sections of the tube. These sections extend above the heat-transfer fluid or metal block. The sample can easily be recentrifuged to the bottom of the tube and a new determination of the boiling point begun. Note that if the bell cavity fills completely during the cooling point of a cycle, it is often difficult to reinitiate the bubble stream without first emptying the entire cavity by overrunning the boiling point.

Observed boiling points for a series of compounds, which boil over a wide range of temperatures, are summarized in Table 4.1.

TABLE 4.1 *Observed Boiling Points (°C)*

COMPOUND	OBSERVED	LITERATURE VALUE	REFERENCE
Methyl iodide	42.5	42.5	a
Isopropyl alcohol	82.3	82.3	b
2,2-Dimethoxypropane	80.0	83.0	c
2-Heptanone	149–150	151.4	d
Cumene	151–153	152.4	e
Mesitylene	163	164.7	f
p-Cymene	175–178	177.1	g
Benzyl alcohol	203	205.3	h
Diphenylmethane	263–265	264.3	i

Note. (Observed values are uncorrected for changes in atmospheric pressure (corrections all estimated to be less than ±0.5 °C).
Source. CRC Handbook of Chemistry and Physics, 78th ed.; CRC Press: Boca Raton, FL, 1997–1998: [a]no. 7518, p. 3–207; [b]no. 103335, p. 3–282; [c]no. 9909, p. 3–271; [d]no. 6510; p. 3–180; [e]no. 1975, p. 3–55; [f]no. 2359, p. 3–66; [g]no. 1190, p. 3–56; [h]no. 1836, p. 3–52; [i]no. 1961, p. 3–55.

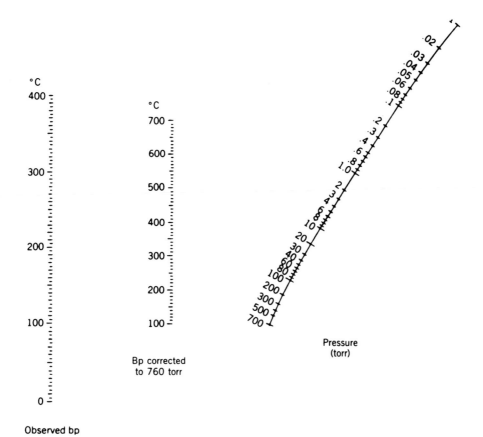

Figure 4.4 Pressure-temperature nomograph.

Materials that are thermally stable at their boiling point will give identical values on repeat determinations. Substances that begin to decompose will give values that slowly drift after the first few measurements. The observation of color and/or viscosity changes, together with a variable boiling point, signal the need for caution in making extended repeat measurements.

Comparison of the boiling points obtained experimentally at various atmospheric pressures with reference boiling points at 760 torr is greatly facilitated by the use of pressure—temperature nomographs such as that shown in Figure 4.4. A straight line from the observed boiling point to the observed pressure will pass through the corrected boiling-point value. These values can be of practical importance when carrying out reduced pressure distillations.

DENSITY

Density, defined as mass per unit volume, is generally expressed as grams per milliliter (g/mL) or grams per cubic centimeter (g/cm^3) for liquids. Accurate nondestructive procedures have been developed for the measurement of this

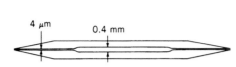

Figure 4.5 Pycnometer of Clemo and McQuillen. (*From Schneider, F. L. Monographien aus dem Gebiete der qualitativen Mikroanalyse, Qualitative Organic Microanalysis, Vol. II; Benedetti-Pichler, A. A., Ed.; Springer: Vienna, 1964.*)

physical constant at the microscale level. A micropycnometer (density meter), developed by Clemo and McQuillen requires approximately 2 μL (Fig. 4.5).[5]

This very accurate device gives the density to three significant figures. The system is self-filling, and the fine capillary ends do not need to be capped while temperature equilibrium is reached or during weighing (the measured values tend to degrade for substances boiling under 100 °C and when room temperatures rise much above 20 °C). In addition, the apparatus must first be tared, filled, and then reweighted on an *analytical* balance. A technique that results in less precise densities (good to about two significant figures), but which is far easier to use, is simply to substitute a 50- or 100-μL syringe for the pycnometer. The method simply requires weighting the syringe before and after filling it to a measured volume as in the conventional technique. With the volume and the weight of the liquid known, the density can be calculated. A further advantage of the syringe technique is that the pycnometer is not limited to a fixed volume. Although much larger samples are required, it is not inconvenient to utilize the entire sample obtained in the reaction for this measurement, since the material can be efficiently recovered from the syringe for additional characterization studies. Because density changes with temperature, these measurements should be obtained at a constant temperature.

An alternative to the syringe method is to use *Drummond Disposable Microcaps* as pycnometers. These precision-bore capillary tubes, calibrated to contain the stated volume from end to end (accuracy ±1%), are available from a number of supply houses.[6] These tubes are filled by capillary action or by suction using a vented rubber bulb (provided). The pipets can be obtained in various sizes, but as with the syringe, volumes of 50, 75, or 100 μL are recommended. When using this method, handle the micropipet with forceps and not with your fingers (it's hot). The empty tube is first *tared*, and then filled and weighed again. The difference in these values is the weight of liquid in the pipet. For convenience, the pipet may be placed in a small container (10-mL beaker or Erlenmeyer flask) when the weighing procedure is carried out.

Two inexpensive micropycnometers can also be easily prepared: The first can be made from a Pasteur pipet as reported by Singh et al.[7] The volume of each individual pycnometer can be varied from 20 to 100 μL, or larger if desired. Values to three significant figures are obtained using an *analytical balance*, because evaporation is generally negligible, and if the pycnometer mouth is small.

[5]Clemo, G. R.; McQuillen, A. *J. Chem. Soc.* **1935,** 1220.

[6]Drummond Disposable Microcaps are available from Arthur H. Thomas Co., Philadelphia, PA 19105; and Sargent-Welch Scientific Co., a VWR company, Skokie, IL.

[7]Singh, M. M.; Szafran, Z.; Pike, R. M. *J. Chem. Educ.* **1993,** *70,* A36; see also Ellefson-Kuehn, J., and Wilcox, C. J. *J. Chem. Educ.* **1994,** *71,* A150; and Singh, M. M.; Pike, R. M.; Szafran, Z. *Microscale and Selected Macroscale Experiments for General and Advanced General Chemistry;* Wiley: New York, 1995.

The second pycnometer, by Pasto and co-workers, is made from a melting-point capillary tube.[8] In both of these techniques, the volume of the pycnometer must be determined. The procedure to determine the density involves the following steps. The empty micropycnometer is tared on an analytical balance, filled with the liquid in question, and reweighed (the difference in weights is the weight of the liquid). The sample is removed and the pycnometer is rinsed with acetone and dried. It is then filled with distilled water and reweighed. From the known[9] density of water at the given temperature the volume of water can be determined and thus the volume of the pycnometer. The volume of the original liquid sample also equals this value. The weight and volume of the sample are used to calculate its density.

REFRACTIVE INDEX

A beam of light appears to bend as it passes from one medium to another. For example, when viewed from above an oar looks bent at the point where it enters the water. This effect is a consequence of the refraction of light. It results from the change in velocity of the radiation at the interface of the media, and the angle of refraction, ϕ'. It is related to the velocity change as follows (see Fig. 4.6):

$$\frac{\sin \phi}{\sin \phi'} = \frac{\text{velocity in vacuum}}{\text{velocity in sample}} = n \text{ (refractive index)}$$

where ϕ is the angle of incidence between the beam of light and the interface. Because the velocity of light in a medium must be less than that in a vacuum, the index of refraction n will always be greater than 1. In practice, n is taken as the ratio of the velocity of light in air relative to the medium being measured.

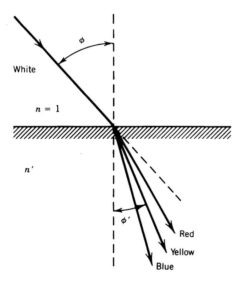

Figure 4.6 Upon refraction, white light is spread out into a spectrum. This is called dispersion.

[8]Pasto, D.; Johnson, C.; Miller, M. *Experiments and Techniques in Organic Chemistry*; Prentice-Hall: Englewood Cliffs, NJ, 1992.

[9]Values for the density of water at various temperatures can be found in the *CRC Handbook of Chemistry and Physics*.

The refractive index is also wavelength dependent. The wavelength dependence gives rise to the effect of dispersion or the spreading of white light into its component colors. When we measure n, therefore, we must specify the wavelength at which the measurement is made. The standard wavelength for refractive index determinations has become the (yellow) sodium 589-nm emission, the sodium D line. Sodium, unfortunately, is a poor choice of wavelength for these measurements with organic substances. In the past, however, the sodium lamp represented one of the easiest-to-obtain monochromatic sources of light and thus it became widely used. Because the density of the medium is sensitive to temperature, the velocity of radiation also changes with temperature, so refractive index measurements must be made at constant temperatures. Many values in the literature are reported at 20 °C. The refractive index can be measured optically quite accurately to four decimal places. Because this measurement is particularly sensitive to the presence of impurities, the refractive index can be a valuable physical constant for tracking the purification of liquid samples.

For example, the measurement is reported as

$$n_D^{20} = 1.4628$$

Procedure

In the Abbe-3L refractometer (Fig. 4.7), white light is used as the source, but compensating prisms give indexes for the D line. This refractometer is commonly used in many organic laboratories. Samples ($\sim$10 μL) are applied between the horizontal surfaces of a pair of hinged prisms (Fig. 4.8). A sampling procedure recently developed by Ronald[10] significantly reduces the amount of

Figure 4.7 Abbe-3L refractometer. *(Courtesy of Milton Roy Co., Rochester, NY.)*

---

[10]Ronald, B. P., Department of Chemistry, Idaho State University, Pocatello, ID (personal communication).

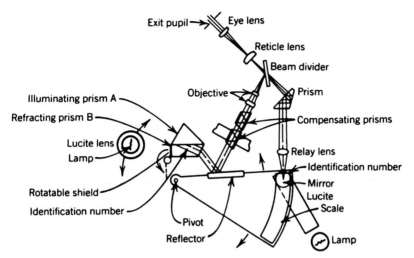

Figure 4.8 Diagram of a typical refractometer. (*Courtesy of Milton Roy Co., Rochester, NY.*)

sample required and allows accurate measurements of highly volatile materials. The technique involves placing a small precut 6-mm disk of good-quality lens paper at the center of the bottom prism. The sample is loaded onto the disk with a micro Pasteur pipet or microliter syringe (see Table 4.2).

Caution

Do not touch the refractometer prisms with the Pasteur pipet or syringe tip. The prisms can be easily and permanently marred or scratched, and the refractometer will then give erroneous results.

The refractometer is adjusted so that the field of view has a well-defined light and dark split image (see your instrument manual for the correct routine for making adjustments on your particular refractometer).

TABLE 4.2 *Refractive Index Measurements Utilizing Lens Paper Disk Technique*

SUBSTANCE	T (°C)	n^t (normal) 100μL	n^t (microdisk) 2–4μL
Water	24.5	1.3224	1.3226
Diethyl ether	24	1.3508	1.3505
Chlorobenzene	24.5	1.5225	1.5219
Iodobenzene	24.5	1.6151	1.6151

When using the refractometer, always clean the prisms with alcohol and lens paper before and after use. Record the temperature at which the reading is taken. A reasonably good extrapolation of temperature effects can be obtained by assuming that the index of refraction changes 0.0004 unit per degree Celsius, and that it varies inversely with temperature.

II. Solids

MELTING POINTS

In general, the crystalline lattice forces holding organic solids together are distributed over a relatively narrow energy range. The melting points of organic compounds, therefore, are usually relatively sharp, that is, less than 2 °C. The range and maximum temperature of the melting point, however, are very sensitive to impurities. Small amounts of sample contamination by soluble impurities nearly always will result in melting-point depressions.

The drop in melting point is usually accompanied by an expansion of the melting-point range. Thus, in addition to the melting point acting as a useful guide in identification, it also can be a particularly effective indication of sample purity.

Procedure

In the microscale laboratory, two different types of melting-point determinations are carried out: (1) simple capillary melting points and (2) evacuated melting points.

Simple capillary melting point

Because the microscale laboratory utilizes the Thomas-Hoover Uni-Melt apparatus or a similar system for determining boiling points, melting points are conveniently obtained on the same apparatus. The Uni-Melt system utilizes an electrically heated and stirred silicone oil bath. The temperature readings require no correction in this case because the depth of immersion is held constant. (This assumes, of course, that the thermometer is calibrated to the operational immersion depth.) Melting points are determined in the same capillaries as boiling points. The capillary is loaded by introducing about 1 mg of material into the open end. The sample is then tightly packed (~2 mm) into the closed end by dropping the capillary down a length of glass tubing held vertically to the bench top. The melting-point tube is then ready for mounting on the metal stage, which is immersed in the silicone oil bath of the apparatus. If the melting point of the substance is expected to occur in a certain range, the temperature can be rapidly raised to ~20 °C below the expected value. At that point, the temperature rise should be adjusted to a maximum of 2 °C/min, which is the standard rate of change at which the reference determinations are obtained. The melting-point range is recorded from the temperature at which the first drop of liquid forms (point *e* in Fig. 4.9) to that at which the last crystal melts (point *m* in Fig. 4.9).

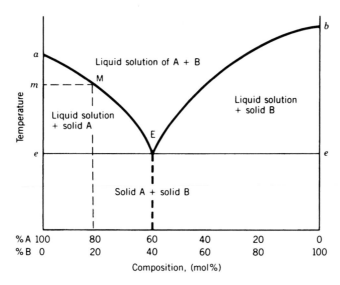

Figure 4.9 Melting point composition diagram for the binary mixture, $A + B$. In this diagram, a is the melting point of the solid A, b of solid B, e of eutectic mixture E, and m of the 80% A:20% B mixture, M.

Evacuated melting points

Many organic compounds begin to decompose at their melting points. This decomposition often begins as the melting point is approached and may adversely affect the values measured. The decomposition can be invariably traced to reaction with oxygen at elevated temperatures. If the melting point is obtained in an evacuated tube, therefore, much more accurate melting points can be obtained. These more reliable values arise not only from increased sample stability, but because several repeat determinations can often be made on the same sample. The multiple measurements then may be averaged to provide more accurate data.

Evacuated melting points are quickly and easily obtained with a little practice. The procedure is as follows: Shorten the capillary portion of a Pasteur pipet to approximately the same length as a normal melting-point tube (Fig. 4.10a). Seal the capillary end by rotating in a microburner flame. Touch the pipet only to the very edge of the flame, and keep the large end at an angle below the end being sealed (Fig. 4.10b). This technique will prevent water from the flame being carried into the tube, where it will condense in the cooler sections. Then load 1–2 mg of sample into the drawn section of the pipet with a microspatula (Fig. 4.10c). Tap the pipet gently to seat the solid powder as far down the capillary as it can be worked (Fig. 4.10d). Then push the majority of the sample part way down the capillary with the same diameter copper wire that you used to seat the cotton plug in constructing the Pasteur filter pipet (Fig. 4.10e). Next, connect the pipet with a piece of vacuum tubing to a mechanical high-vacuum pump. Turn on the vacuum and evacuate the pipet for 30 s (Fig. 4.10f). With a microburner, gently warm the surface of the capillary tubing just below the drawn section. On warming, the remaining fragments of the sample (the majority of which has been forced farther down in the tube) will sublime in either direction away from the hot section. Once the traces of sample have been "chased" away, the heating is increased, and the capillary tube is collapsed, fused, and separated from the shank. The shank remains connected to the vacuum system (Fig. 4.10g). The vacuum system is then vented and the shank is discarded. The sample is tightly packed into the initially sealed

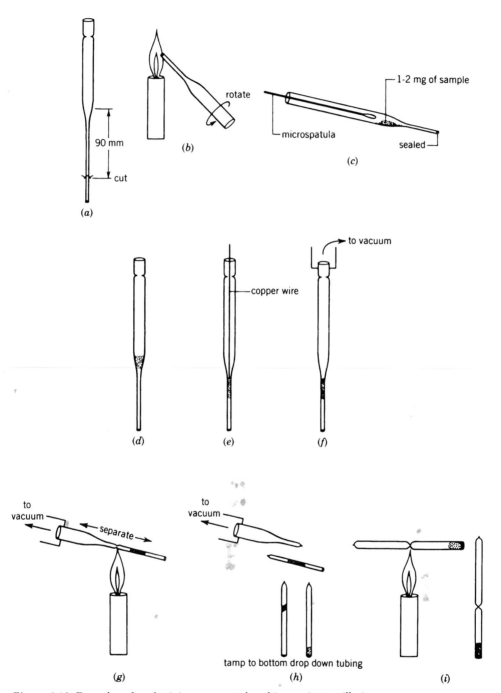

Figure 4.10 Procedure for obtaining evacuated melting-point capillaries.

end of the evacuated capillary by dropping it down a section of glass tubing, as in the case of packing open melting-point samples. After the sample is packed (~2 mm in length, see Fig. 4.10*h*), a section of the evacuated capillary about 10–15 mm above the sample is once more gently heated and collapsed by the microburner flame (Fig. 4.10*i*).

This procedure is required to trap the sample below the surface of the heated silicone oil in the melting-point bath, and thus avoid sublimation up the tube to cooler sections during measurement of the melting point. The operation is a little tricky and should be practiced a few times. It is very important that the tubing completely fuse. Now the sample is ready to be placed in the melting-point apparatus. The procedure beyond this point is the same as in the open capillary case, except that after the sample melts, it can be cooled, allowed to crystallize, and remelted several times, and the average value of the range reported. If these values begin to drift downward, the sample can be considered to be decomposing even under evacuated, deoxygenated conditions. In this case the first value observed should be recorded as the melting point and decomposition noted (mp xx dec, where dec = decompose).

Mixture melting point

Additional information can often be extracted from the sensitivity of the melting point to the presence of impurities. Where two different substances possess identical melting points (not uncommon), it would be impossible to identify an unknown sample as either material based on the melting point alone. If reference standards of the two compounds are available, however, then mixtures of the unknown and the two standards can be prepared. it is important to prepare several mixtures of varying concentrations for melting-point comparisons, since the point of maximum depression need not occur on the phase diagram at the 50:50 ratio (see Fig. 4.9). In samples that do not exhibit any decomposition at the melting point, the prepared mixtures should be first heated until a homogeneous melt is obtained. Each is then cooled and ground to a fine powder, and the definitive melting point is obtained on the ground sample. The melting points of the unknown and the mixed samples should be obtained simultaneously (the Uni-Melt stage will accept up to seven capillaries at one time). This is desirable because all the samples will then be heated at the same rate. The unknown sample and the mixture of the unknown with the correct reference will have identical values, but the mixture of the reference with a different substance will give a depressed melting point. This procedure is the classical step to positive identification of crystalline solids.

Mixtures of two different compounds only rarely fail to exhibit mixture melting-point depression, but it can happen. Some mixtures may not show a depression or show only a very small one, due to eutectic or compound formation. Elevation of the melting point has also been observed. Therefore, if mixture melting-point data are used for identification purposes, comparison of other physical constants or spectroscopic data is advocated to establish identity beyond any reasonable doubt.

QUESTIONS

4-1. Room temperature is recorded when a density determination for a given substance is performed in the laboratory. Why?

4-2. Describe how you would determine the melting point of a substance that sublimes before it melts.

4-3. What are the benefits of determining the refractive index of a liquid?

4-4. In the microscale method of determining boiling points, one heats the liquid until a steady stream of bubbles is observed coming out of the bell. The temperature is then lowered and the boiling point is read just as the bubbles stop. Why is this technique preferable to measuring the boiling point when the bubbles first start to appear?

4-5. The boiling point of a liquid was recorded as 110 °C at 1.0 torr. What is the boiling point corrected to 760 torr?

5

Microscale

Laboratory Techniques

This chapter introduces the microscale organic laboratory techniques that must be mastered to be successful when working at this scale. Detailed discussions are given for each individual experimental technique. *New to this edition is our accompanying website, www.wiley.com/college/MTOL2. To streamline our treatment of the laboratory, we have removed a considerable quantity of basic reference material from the text and placed it in easily accessible form on our website. The icon ✹ is used throughout the text to indicate website material that will be of interest to the user. We hope this change will make the more important aspects of the basic text easier to access and will speed along your laboratory work.*

One of the principal hurdles in dealing with experimental chemistry is the isolation of pure materials. Characterization (identification) of a substance almost always requires a pure sample of the material. This is a particularly difficult demand of organic chemistry because most organic reactions generate several products. We are generally satisfied if the desired product is the major component of the mixture obtained. This chapter places a heavy emphasis on separation techniques.

TECHNIQUE 1

Microscale Separation of Liquids by Preparative Gas Chromatography

Technique 1 begins the discussion of the resolution (separation) of microliter quantities of liquid mixtures via preparative gas chromatography. Techniques 2 and 3 deal with semimicro adaptations of classical distillation routines that focus on the separation of liquid mixtures involving one to several milliliters of material.

Chromatography methods revolutionized experimental organic chemistry. These methods are by far the most powerful of the techniques for separating mixtures and isolating pure substances, either solids or liquids. Chromatography is the resolution (separation) of a multicomponent mixture (several hundred components in some cases) by distribution between two phases, one stationary and one mobile. The various methods of chromatography are categorized by the phases involved: column, thin–layer, and paper (all solid–liquid chromatography); partition (liquid–liquid chromatography); and vapor phase (gas–liquid chromatography, or simply gas chromatography). The principal mechanism these separations depend on is differential solubility, or adsorbtivity, of the mixture components in the two phases involved. That is, the components must exhibit different partition coefficients (see also Technique 4 for a detailed discussion of partition coefficients).

Gas chromatography (GC, sometimes called vapor-phase chromatography) is an extraordinarily powerful technique for separating mixtures of organic compounds. The stationary phase in GC is a high-boiling liquid and the mobile phase is a gas (the carrier gas). Gas chromatography can separate mixtures far better than distillation techniques can (see Technique 2 discussion).

Preparative GC separations, which involve perhaps 5–100 µL of material, require relatively simple instrumentation but sacrifice resolution for the ability to separate larger amounts of material.

Analytical GC separations require tiny amounts of material (often 0.1 µL of a very dilute solution), and can separate incredibly complex mixtures. The need to work with small quantities of materials in analytical GC separations is an advantage at the microscale level. Analytical GC is used to analyze distillation samples in Examples [3A–3B], pp. 83–87.

GC instrumentation

GC instrumentation can range from straightforward and relatively simple systems to systems with complex, highly automated, and relatively expensive components. A diagram of a common and simple GC typically used in an instructional laboratory is shown in Figure 5.1.

Injection Port   The analysis begins in a heated injection port. The sample mixture is introduced by syringe through a septum into the high-temperature chamber (injection port) through which the inert carrier gas (the mobile phase) is flowing. Helium and nitrogen are common carrier gases. The solubility of the sample in the carrier gas depends mostly on the vapor pressure of the substances in the sample. Heating the injection port helps to ensure the vaporization of less volatile samples. There are two major constraints on GC: The sample must be stable at the temperature required to cause vaporization, and the sample must have sufficient vapor pressure to be completely soluble in the carrier gas at the column operating temperatures.

Note

When injecting a sample, always position your thumb or finger over the syringe plunger. This prevents a blow-back of the sample by the carrier gas pressure in the injection port.

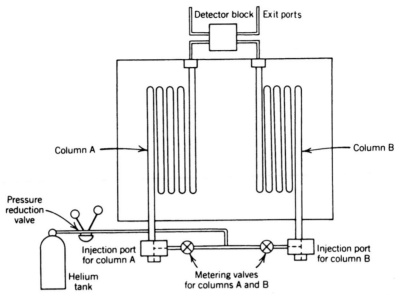

Figure 5.1 Block diagram of a dual-column gas chromatograph showing essential parts. *(Courtesy of GOW-MAC Instrument Co., Bound Brook, NJ.)*

Column The vaporized mixture is swept by the carrier gas from the injection port onto the column. Bringing the sample mixture into intimate contact with the column begins the separation process. The stationary liquid phase in which the sample will partially dissolve is physically and/or chemically bonded to inert packing material (often called the support) in the column. Gas-chromatographic columns are available from manufacturers in a variety of sizes and shapes. In the diagram of the GOW-MAC instrument (Fig. 5.1), two parallel coiled columns are mounted in an oven. Considerable oven space can be saved and better temperature regulation achieved if the columns are coiled. Temperature regulation is particularly important, because column resolution degrades rapidly if the entire column is not at the same temperature. Most liquid mixtures need a column heated above ambient temperatures to achieve the vapor pressure the separation requires.

The mixture is separated as the carrier gas sweeps the sample through the column. Columns are usually made from stainless steel, glass, or fused silica. The diameter and length of the column are critical factors in separating the sample mixture.

Packed Columns In packed columns the liquid (stationary) phase in contact with the sample contained in the mobile gas phase is maximized by coating a finely divided inert support with the nonvolatile liquid. The coated support is carefully packed into the column so as not to develop empty spaces. Packed columns are usually 1/4 or 1/8 inch in diameter and range from 4 to 12 feet in length. These columns are particularly useful in the microscale laboratory, since they can be used for both analytical and preparative GC. Simple mixtures of 20–80 µL of material can often be separated into their pure components and collected at the exit port of the detector. Smaller samples (0.2–2.0-µL range) will exhibit better separation.

Capillary Columns Capillary columns have no packing; the liquid phase is simply applied directly to the walls of the column. These columns are referred to as wall-coated, open-tubular (WCOT) columns. The reduction in surface area (compared to packed columns) is compensated for by tiny column diameters (perhaps 0.1 mm) and impressive lengths (100 m is not uncommon). Capillary columns are the most powerful columns used for analytical separations. Mixtures of several hundred compounds can be completely resolved on a capillary GC column. These columns require a more sophisticated and expensive chromatography instrument. Capillary columns, because of their tiny diameters, can accommodate only very small samples, perhaps 0.1 μL or less of a dilute solution. Capillary columns cannot be used for preparative separations.

Liquid Phase Once the sample is introduced on the column (in the carrier gas), it will undergo partition into the liquid phase. The choice of the liquid phase is particularly important because it directly affects the relative distribution coefficients.

In general, the stationary liquid phase controls the partitioning of the sample by two criteria. First, if little or no interaction occurs between the sample components and the stationary phase, the boiling point of the materials will determine the order of elution. Under these conditions, the highest boiling species will be the last to elute. Second, the functional groups of the components may interact directly with the stationary phase to establish different partition coefficients. Elution then depends on the particular binding properties of the sample components.

Some typical materials used as stationary phases are shown next.

NAME	STATIONARY PHASE	MAXIMUM TEMPERATURE (°C)	MECHANISM OF INTERACTION
Silicone oil DC710, etc.	$R_3[OSiR_2]_nOSiR_3$	250	According to boiling point
Polyethylene glycol (Carbowax®)	$HO[CH_2CH_2O]_nCH_2CH_2OH$	150	Relatively selective toward polar components
Diisodecyl phthalate	$o\text{-}C_6H_4[CO_2\text{-isodecyl}]_2$	175	According to boiling point

Oven Temperature The temperature of the column will also affect the separation. In general, the elution time of a sample will decrease as the temperature is increased. That is, retention times are shorter at higher temperatures. Higher boiling components tend to undergo diffusion broadening at low column temperatures because of the increase in retention times. If the oven temperature is too high, however, equilibrium partitioning of the sample with the stationary phase will not be established. Then the components of the mixture may elute together or be incompletely separated. Programmed oven temperature increases can speed up elution of the higher boiling components, but suppress peak broadening and therefore increase resolution. Temperature-programming capabilities require more sophisticated ovens and controllers.

Flow Rate   The flow rate of the carrier gas is another important parameter. The rate must be slow enough to allow equilibration between the phases, but fast enough to ensure that diffusion will not defeat the separation of the components.

Column Length   As noted, column length is an important factor in separation performance. As in distillations, column efficiency is directly proportional to column height, which determines the number of evaporation-condensation cycles. In a similar manner, increasing the length of a GC column allows more partition cycles to occur. Difficult-to-separate mixtures, such as the xylenes (very similar boiling points: *o*-xylene, 144.4 °C; *m*-xylene, 139.1 °C; and *p*-xylene, 138.3 °C), have a better chance of being separated on longer columns. In fact, both GC and distillation resolution data are described using the same term, *theoretical plates* (see Technique 3, Examples [2B–2C], pp. 83–87.

Detector and Exit Port   A successfully separated mixture will elute as its individual components at the instrument's exit port (also temperature controlled). To monitor the exiting vapors, a detector is placed in the gas stream just before the exit port (Fig. 5.1). After passing through the detector, the carrier gas and the separated sample components are then vented.

One widely used detector is the nondestructive, thermal conductivity detector, sometimes called a hot-wire detector. A heated wire in the gas stream changes its electrical resistance when a substance dilutes the carrier gas and thus changes its thermal conductivity. Helium has a higher thermal conductivity than most organic substances. When substances other than helium are present, the conductivity of the gas stream changes, which changes the resistance of the heated wire. The change in resistance is measured by comparing it to a reference detector mounted in a second (parallel) gas stream (Wheatstone bridge). The resulting electrical signal is plotted on a chart recorder, where the horizontal axis is time and the vertical axis is the magnitude of the resistance difference. The plot of resistance difference versus time is referred to as the gas *chromatogram*. The retention time (t_R) is defined as the time from sample injection to the time of maximum peak intensity. The baseline width (W_b) of a peak is defined as the distance between two points where tangents to the points of infection cross the baseline (Fig. 5.2).

Capillary GC systems, and other GC systems used only for analytical separations, often use a flame-ionization detector (FID). In a flame-ionization detector, the gas eluting from the GC column is mixed with air (or oxygen) and hydrogen, and burned. The conductivity of the resulting flame is measured; it changes with the ionic content of the flame, which is proportional to the amount of carbon (from organic material) in the flame. The advantage of an FID is its high sensitivity; amounts of less than a microgram are easily detected. Its disadvantage is that it destroys (burns) the material it detects.

Theoretical plates   It is possible to estimate the number of theoretical plates (directly related to the number of distribution cycles) present in a column for a particular substance. The parameters are given in the relationship[1]

$$ n = 16 \left(\frac{t_R}{W_b} \right)^2 $$

---

[1]Berg, E. W. *Physical and Chemical Methods of Separation*; McGraw-Hill: New York, 1963, p. 111.

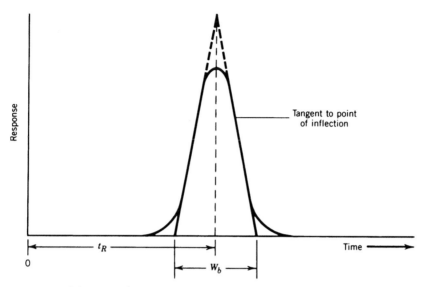

Figure 5.2 Schematic chromatogram.

where the units of retention time (t_R) and baseline width (W_b) are identical (minutes, seconds, or centimeters). As in distillation columns, the larger the number of theoretical plates, n, the higher the resolution of the column and the better the separation.

The efficiency of a system may be expressed as the *height-equivalent theoretical plate* (HETP) in centimeters (or inches) per plate. The HETP is related to the number of theoretical plates n by

$$\text{HETP} = \frac{L}{n}$$

where L is the length of the column, usually reported in centimeters. The smaller the HETP, the more efficient the column.

The number of theoretical plates available in fractional distillation columns is limited by column holdup (see Techniques 2A and 2B). Thus, distillations of less than 500 μL are generally not practical. Gas-chromatographic columns, on the other hand, operate most efficiently at the microscale or submicroscale levels, where 500 μL would be an order of magnitude (even 3–8 orders of magnitude in the case of capillary columns) too large.

Fraction Collection Sequential collection of separated materials can be made by attaching suitable sample condensing tubes to the exit port (see Fig. 3.6, p. 20).

Procedure for preparative collection

The collection tube (oven dried until 5 min before use) is attached to the heated exit port by the metal 5/5 ₮ joint. Sample collection is begun 30 s before detection of the expected peak on the recorder (based on previously determined retention values; refer to your local laboratory instructions) and continued until 30 s after the recorder's return to baseline. After the collection tube is detached, the sample is transferred to the 0.1-mL conical GC collection vial. After the collection tube is joined to the vial (preweighed with cap) by the 5/5

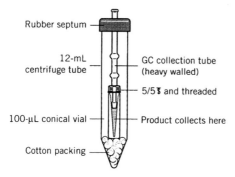

Rubber septum

12-mL centrifuge tube

GC collection tube (heavy walled)

5/5 ꝣ and threaded

100-μL conical vial

Product collects here

Cotton packing

Figure 5.3 Gas chromatographic collection tube and 0.1-mL conical vial.

ꝣ joints, the system is centrifuged to force the sample down into the vial (Fig. 5.3). The collection tube is then removed, and the vial is capped and reweighed.

The efficiency of collection can exceed 90% with most materials, even with relatively low-boiling substances. In the latter case, the collection tube, after attachment to the instrument, is wrapped with a paper tissue. As the (oven-dried) tube is being wrapped, it is also being flushed by the carrier gas, which removes any traces of water condensation. The wrapping is then saturated with liquid nitrogen to cool the collection tube.

Preparative GC in the microscale laboratory often replaces the macroscale purification technique of fractional distillation. Distillation is impractical with less than 500 μL of liquid.

Refer to Examples [1A–1D], pp. 64–71, for further specific experimental details on preparative GC applied to the separation of a number of binary (two-component) mixtures. These are designed as practice examples to give you experience with sample collection.

QUESTIONS

5-1. What is the main barrier to separating liquid mixtures of less than 500 μL by distillation?

5-2. A sample mixture of ethyl benzoate (bp 212 °C) and dodecane (bp 216.2 °C) is injected on two GC columns. Column A has DC710 silicone oil as the stationary phase, and column B uses polyethylene glycol as the stationary phase. Which substance would be certain to elute first from column A, and would the same material be expected to elute first from column B? Which column, A or B, would be expected to give the better separation of these two substances?

5-3. Question 5-2 refers to separating a mixture of two high boiling liquids by gas chromatography. These materials have similar boiling points. List several GC variables and conditions that would make it easier to separate these substances by gas chromatography.

5-4. Capillary GC columns have better resolution than packed columns even though the enormous surface area provided by the packing material is absent in capillary columns. Why?

5-5. Preparative GC requires packed columns. Why is this technique limited to these lower resolution columns?

EXAMPLE [1A]

The Separation of a 25-μL Mixture of Heptanal (bp 153 °C) and Cyclohexanol (bp 160 °C) by Gas Chromatography

Purpose

This example illustrates the separation of a 25-μL mixture, consisting of heptanal and cyclohexanol, into the pure components. The volume of the mixture is approximately that of a single drop, and the materials boil within 7 °C of each other. This mixture would be difficult, if not impossible, to separate by the best distillation techniques available. The purity of the fractions collected from the gas chromatograph (GC) can be assessed by boiling points, refractive indexes, or infrared (IR) spectra.

DISCUSSION

The efficacy of GC separations is highly dependent on the experimental conditions. For example, two sets of experimental data on the heptanal-cyclohexanol mixture are given below to demonstrate the effects of variations in oven temperature on retention times.

In Data Set A, the oven temperature was allowed to rise slowly from 160 to about 170 °C during a series of sample collections. The retention time of heptanal dropped from about 3 min to close to 2 min, whereas the retention time of cyclohexanol was reduced from about 5.5 min to nearly 4 min. The significant decrease in resolution over this series of collections is reflected in the number of theoretical plates calculated, which was over 300 for heptanal and about 500 for cyclohexanol in the first trial, but declined to below 200 for both compounds toward the last run (see Data Set A).

COLLECTION YIELD

Cyclohexanol

Density of cyclohexanol = 0.963 mg/μL.

In 25 μL of 1 : 1 cyclohexanol-heptanal, we have 12.5 μL of cyclohexanol.

Therefore, 12.5 μL × 0.963 mg/μL = 12 mg of cyclohexanol injected.

Percent recovered = (8.3 mg/12.0 mg) × 100 = 69% cyclohexanol collected.

Data Set A

TRIAL NO.	Heptanal				Cyclohexanol			
	RETENTION TIME (min)	BASELINE WIDTH (min)	NUMBER OF THEORETICAL PLATES	RECOVERY (mg)	RETENTION TIME (min)	BASELINE WIDTH (min)	NUMBER OF THEORETICAL PLATES	RECOVERY (mg)
1	3.1	0.7	314	8.0	5.6	1.0	502	8.0
2	2.9	0.7	275	8.0	5.3	1.0	449	8.0
3	3.0	0.7	294	7.0	5.7	1.0	520	8.0
4	2.8	0.7	256	8.0	5.1	1.1	344	8.0
5	2.5	0.6	278	8.0	4.3	1.1	244	9.0
6	2.7	0.5	467	7.0	4.6	1.0	339	10.0
7	2.5	0.6	278	10.0	4.2	1.0	282	8.0
8	2.2	0.5	310	9.0	3.5	1.0	196	8.0
9	1.8	0.5	207	8.0	3.0	1.0	144	8.0
10	2.3	0.7	173	8.0	3.9	1.0	243	8.0
Av	2.5 ± 0.4	0.6 ± 0.09	285 ± 7	8.1 ± 0.8	4.5 ± 0.9	1.0 ± 0.05	326 ± 129	8.3 ± 0.7

Heptanal

> Density of heptanal = 0.850 mg/μL.
>
> Therefore, 12.5 μL × 0.85 mg/μL = 10.6 mg of heptanal injected.
>
> Percent recovered = (8.1 mg/10.6 mg) × 100 = 76% heptanal.

In the Data Set B collections, stable oven temperatures and flow rates were maintained, and the data exhibit excellent reproducibility. Oven temperature was held at 155 °C throughout the sampling process. The retention time of heptanal was observed to be slightly longer than 3 min, with a variance of 6 s, whereas the cyclohexanol retention time was found to be slightly longer than 6 min, with a variance of 12 s. The resolution remained essentially constant throughout the series, and the number of theoretical plates calculated was about 350 for heptanal and about 500 for cyclohexanol (see Data Set B).

COLLECTION YIELD

Cyclohexanol

> Density of cyclohexanol = 0.963 mg/μL.
>
> In 25 μL of 1:1 cyclohexanol-heptanal, there are 12.5 μL of cyclohexanol.
>
> Therefore, 12.5 μL × 0.963 mg/μL = 12 mg of cyclohexanol injected.
>
> Percent recovered = (8.5 mg/12.0 mg) × 100 = 71% cyclohexanol collected.

Heptanal

> Density of heptanal = 0.850 mg/μL.
>
> Therefore, 12.5 μL × 0.85 mg/μL = 10.6 mg of heptanal injected.
>
> Percent recovered = (8.3 mg/10.6 mg) × 100 = 78% heptanal.

The results just described demonstrate that the resolution of GC peaks may be very sensitive to changes in retention time resulting from instability in oven temperatures. Since the number of theoretical plates is related to resolution values, significant degradation in column plate values can occur with variations in oven temperatures. When you compare the time and effort required to obtain a two-plate fractional distillation on a 2-mL mixture (see Technique 2 and Example [2A]) with the speed and ease used to obtain a 500 plate separation on 12.5 μL of cyclohexanol in this example, it is hard not to be impressed with the enormous power of this technique.

Data Set B

TRIAL NO.	Heptanal				Cyclohexanol			
	RETENTION TIME (min)	BASELINE WIDTH (min)	NUMBER OF THEORETICAL PLATES	RECOVERY (mg)	RETENTION TIME (min)	BASELINE WIDTH (min)	NUMBER OF THEORETICAL PLATES	RECOVERY (mg)
1	3.5	0.7	400	8.0	6.6	1.1	576	8.0
2	3.2	0.7	334	9.0	6.0	1.1	476	7.0
3	3.5	0.7	400	7.0	6.6	1.2	484	10.0
4	3.2	0.7	334	9.0	6.1	1.0	595	9.0
5	3.1	0.6	427	8.0	6.0	1.1	476	8.0
6	3.2	0.7	334	9.0	6.0	1.1	476	9.0
7	3.3	0.8	272	9.0	6.1	1.1	492	8.0
8	3.1	0.7	313	8.0	6.0	1.1	476	10.0
9	3.2	0.7	334	8.0	6.1	1.1	492	8.0
10	3.2	0.7	334	8.0	6.2	1.1	508	8.0
Av.	3.2 ± 0.1	0.7 ± 0.05	348 ± 47	8.3 ± 0.7	6.2 ± 0.2	1.1 ± 0.05	505 ± 44	8.5 ± 1.0

COMPONENTS

Cyclohexanol   Heptanal

EXPERIMENTAL PROCEDURE

Estimated time for the experiment: 2.0 h.

Physical Properties of Components

COMPOUND	MW	AMOUNT	bp (°C)	DENSITY (d)	n_D
Heptanal	114.19	12.5 μL	153	0.85	1.4113
Cyclohexanol	100.16	12.5 μL	160	0.96	1.4641

Reagents and equipment

The procedure involves injecting a 25-μL mixture of heptanal-cyclohexanol 1:1 (v/v) into a 1/4-in. × 8-ft stainless-steel column packed with 10% Carbowax 80/100 20M PAW-DMS. Experimental conditions (GOW-MAC series No. 350) are: He flow rate, 50 mL/min; chart speed, 1 cm/min; oven temperature, 155 °C.

Procedure for preparative collection

The liquid effluents are collected in an uncooled, 4-mm-diameter collection tube (double reservoirs; overall tube length 40–50 mm, see Fig. 5.3)

 The collection tube (oven dried until 5 min before use) is attached to the heated exit port by the 5/5 Ŧ joint. Sample collection is initiated 0.5 min prior to detection on the recorder of the expected peak (time based on previously determined retention values)[2] and continued until 0.5 min following return to baseline. After the collection tube is detached, the sample is transferred to the 0.1- mL conical GC collection vial. The transfer is facilitated by the 5/5 Ŧ joint on the conical vial. After the collection tube is joined to the vial (preweighed with stopper), the system is centrifuged (see Fig. 5.3). The collection tube is then removed and the vial is stoppered and reweighed.

---

[2]Refer to your local laboratory instructions.

Characterization

Calculate the percent recovery. These amounts should range between 7 and 10 mg. Determine the boiling point of each fraction and obtain the refractive index or IR spectrum, if time permits. These latter measurements will require most, if not all, of the sample not used for boiling-point determination.

Assess the purity and efficiency of the separation from your tabulated data and the GC chromatogram.

Alternative mixture pairs for preparative collection

(all mixtures are 1:1 v/v)

EXAMPLE: [1B]
Mixture

Separation of a 40-μL mixture of[3]
(1S)-(−)-α-pinene (bp 156 °C, n_D = 1.4650, d = 0.855)
(1S)-(−)-β-pinene (bp 165 °C, n_D = 1.4782, d = 0.859)

COMPONENTS

(1S)-(−)-α-Pinene   (1S)-(−)-β-Pinene

Chromatographic parameters
A 40-μL injection
Flow rate: 50 mL/min
Column temperature: 120 °C
Column: 20% Carbowax
Elution time
α-Pinene: ~8 min
β-Pinene: ~12 min
Average recovery
α-Pinene: 8.3 μL
β-Pinene: 10.6 μL

---

[3]Refractive index at D line of sodium = n_D and density = d.

EXAMPLE: [1C]
Mixture

Separation of a 40-μL mixture of
   2-Heptanone (bp 149–150 °C, n_D = 1.4085, d = 0.820)
   Cyclohexanol (bp 160–161 °C, n_D = 1.4641, d = 0.963)

COMPONENTS

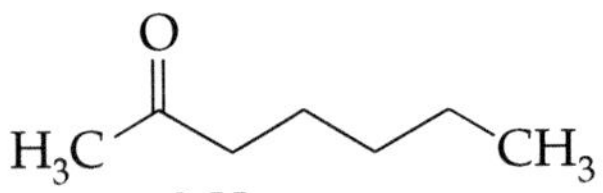

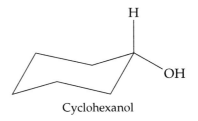

2-Heptanone   Cyclohexanol

Chromatographic Parameters

A 40-μL injection

Flow rate: 50 mL/min

Column temperature: 145 °C

Column: 20% Carbowax

Elution time
   2-Heptanone: ~5.5 min
   Cyclohexanol: ~10.0 min

Average recovery
   2-Heptanone: 8.1 μL (41%)
   Cyclohexanol: 11.4 μL (57%)

EXAMPLE [1D]
Mixture

Separation of a 40-μL mixture of
   d-Limonene (bp 175–176 °C, n_D = 1.4743, d = 0.8402)
   Cyclohexyl acetate (bp 173 °C, n_D = 1.4401, d = 0.9698)

COMPONENTS

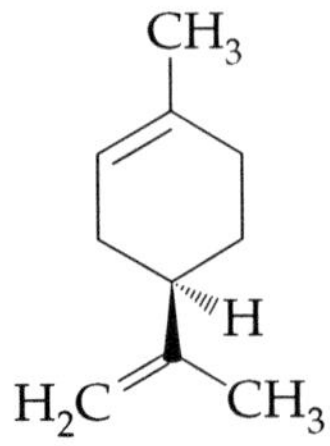

d-Limonene 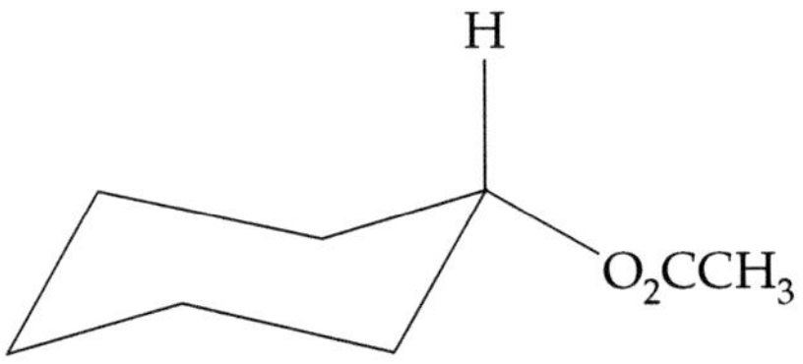   Cyclohexyl acetate

Chromatographic Parameters

A 40-µL injection

Flow rate: 50 mL/min

Column temperature: 170 °C

Column: 20% Carbowax

Elution time
  d-Limonene: ~5.5 min
  Cyclohexyl acetate: ~7.5 min

Average recovery
  d-Limonene: 8.7 µL (44%)
  Cyclohexyl acetate: 10.0 µL (50%)

QUESTIONS

5-6. Based on the data presented in the Data Set A chromatographic separation, can you explain why there is such a steep decline in column efficiency with temperature change?

5-7. Consider the following gas chromatogram for a mixture of analytes X and Y:
  (a) Calculate the number of theoretical plates for the column in reference to the peaks of each component (X and Y).
  (b) If the column is 12 m long, calculate the height equivalent theoretical plate (HETP (in plates per cm) for this column.

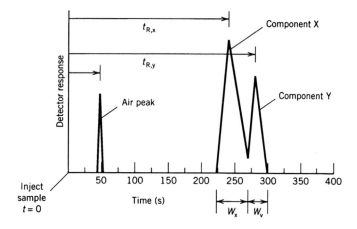

5-8. The number of theoretical plates a column has is important, but the crucial factor is the ability to separate two or more substances. That is, how well resolved are the peaks? The resolution of two peaks depends not only on how far apart they are (t_R), but also on the peak width (W). Baseline resolution (R) is defined by the following equation:

$$R = \frac{2\ \Delta t_R}{W_X + W_Y}$$

Because of the tailing of most species on the column, a value of 1.5 is required to give baseline resolution.

(a) Calculate the resolution for the peaks in Question 5-7.

(b) Do you think a quantitative separation of the mixture is possible based on your answer?

(c) Has baseline resolution been achieved?

5-9. Discuss at least two techniques you might use to increase the resolution of the column in Question 5-8 (without changing the column).

5-10. Retention times for several organic compounds separated on a GC column are given below:

COMPOUND	t_R (s)
Air	75
Pentane	190
Heptane	350
2-Pentene	275

(a) Calculate the relative retention of 2-pentene with respect to pentane.

(b) Calculate the relative retention of heptane with respect to pentane.

BIBLIOGRAPHY

Selected references on gas chromatography:

ETTRE, L. S.; HINSHAW, J. V. *Basic Relationships of Gas Chromatography;* Advanstar Communications: Cleveland, OH, 1993.

GROB, R. L., Ed., *Modern Practices of Gas Chromatography;* Wiley: New York, 1995.

JENNINGS, W., Ed. *Aanalytical Gas Chromatography;* Academic Press: New York, 1997.

McNAIR, H. M.; MILLER, J. M. *Basic Gas Chromatography;* Wiley: New York, 1997.

TECHNIQUES 2 AND 3
Distillation

Distillation is the process of heating a liquid to the boiling point, condensing the heated vapor by cooling, and returning either a portion of, or none of, the condensed vapors to the distillation vessel. Distillation differs from reflux (see p. 25) only in that at least some of the condensate is removed from the boiling system. Distillations in which a fraction of the condensed vapors are returned to the boiling system are often referred to as being under "partial reflux." Two types of distillations will be described under the headings Techniques 2 and 3. Students are encouraged to refer to and study the more detailed discussion of distillation theory, 🌐. In the website discussion, the theory of steam distillation is covered, including an example, Example [2BW], of steam distillation of natural products. There are times when ordinary distillation may not be feasible for the separation of a liquid from dissolved impurities. The compound of interest may boil at a high temperature that is difficult to control with a simple apparatus, or it may tend to decompose or oxidize at high temperatures. If the compound is only sparingly soluble in water and any small amount of water can be removed with a drying agent, steam distillation may be the technique of choice. Compounds that are immiscible in water have very large positive deviations from Raoult's law. Therefore, the boiling temperature is generally lower than that of water and the compound. Included in the website discussion of fractional distillation theory are details for carrying out reduced-pressure distillations, in addition to an example, Example [3BW].

TECHNIQUE 2
Simple Distillation at the Semimicroscale Level

Simple distillation involves the use of the distillation process to separate a liquid from minor components that are nonvolatile, or that have boiling points at least 30–40 °C above that of the major component. A typical setup for a macroscale distillation of this type is shown in Figure 5.4. At the microscale level, working with volumes smaller than 500 µL, GC techniques (see Technique 1) have replaced conventional microdistillation processes.[4] Semimicroscale simple distillation is an effective separation technique for volumes in the range of

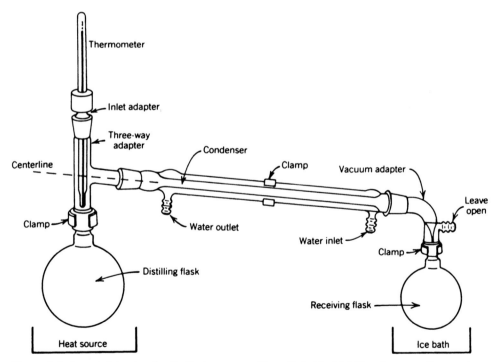

Figure 5.4 A complete simple distillation setup. (*From Zubrick, J. W. The Organic Chem Lab Survival Manual, 4th ed.; Wiley: New York, 1997. Reprinted by permission of John Wiley & Sons, Inc., New York.*)

[4]Schneider, F. L. In *Monographien aus dem Gebiete der qualitatavin Mikroanalyse*, Vol. II: *Qualitative Organic Microanalysis*; Benedetti–Pichler, A. A., Ed.; Springer-Verlag: Vienna, 1964; p. 31.

0.5–2 mL. Apparatus that achieve effective separation of mixture samples in this range have been developed. One of the most significant of these designs is the classic Hickman still, shown in Figure 5.5. This still can be used in several ways: for purifying solvents, carrying out reactions, and concentrating solutions.

In a distillation where liquid is to be separated from a nonvolatile solute, the vapor pressure of the liquid is lowered by the presence of the solute, but the vapor phase consists of only one component. Thus, except for the incidental transfer of nonvolatile material by splashing, the material condensed should consist only of the volatile component.

We can understand what is going on in a simple distillation of two volatile components by referring to the phase diagrams shown in Figures 5.6 and 5.7. Figure 5.6 is the phase diagram for hexane and toluene (see Example [2A], p. 76. The boiling points of these liquids are separated by 42 °C. Figure 5.7 is the phase diagram for methylcyclohexane and toluene. Here the boiling points are separated by only 9.7 °C.

Imagine a simple distillation of the hexane—toluene pair in which the liquid in the pot is 50% hexane. In Figure 5.6, when the liquid reaches 80.8 °C it will be in equilibrium with vapor having a composition of 77% hexane. This result is indicated by the line *A—B*. If this vapor is condensed to a liquid of the same composition, as shown by line *B—C*, we will have achieved a significant enrichment of the condensate with respect to hexane. This change in composition is referred to as a simple distillation. The process of evaporation and condensation is achieved by the theoretical construct known as a *theoretical plate*. When this distillation is actually done with a Hickman still, some of the

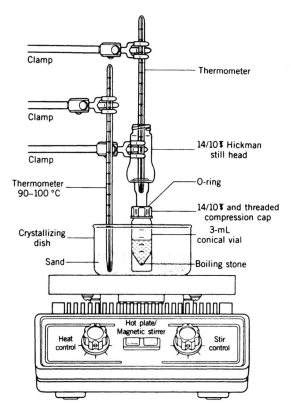

Figure 5.5 Hickman still (14/10 ℥ with conical vial [3 mL]).

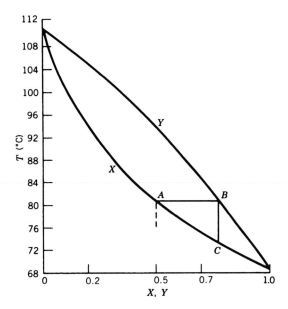

Figure 5.6 Temperature as a function of liquid composition (*X*) and vapor composition (*Y*): hexane and toluene.

mixture will go through one evaporation and condensation cycle, some will go through two of these cycles, and some may be splashed more directly into the collar. A resolution (separation) of between one and two theoretical plates is generally obtained.

Referring to Figure 5.7, if we consider the same process for a 50% mixture of methylcyclohexane and toluene, the methylcyclohexane composition will increase to 58% for a distillation with one theoretical plate. Simple distillation may thus provide adequate enrichment of the MVC (more volatile constituent) if the boiling points of the two liquids are reasonably well separated, as they are for hexane and toluene. If the boiling points are close together, as they are for methylcyclohexane and toluene, the simple distillation will not provide much enrichment.

As we continue the distillation process and remove some of the MVC by condensing it, the residue in the heated flask becomes less rich in the MVC. This means that the next few drops of condensate will be less rich in the MVC. As the distillation is continued, the condensate becomes less and less rich in the MVC.

We can improve on simple distillation by repeating the process. For example, we could collect the condensate until about one-third is obtained. Then we could collect a second one-third aliquot in a separate container. Our original

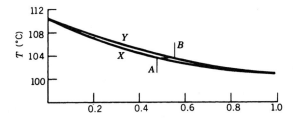

Figure 5.7 Temperature as a function of liquid composition (*X*) and vapor composition (*Y*): methylcyclohexane and toluene.

mixture would then be separated into three fractions. The first third would be richest in the MVC, and the final third (the fraction remaining in the distillation pot) would be the richest in the least volatile component. If the MVC were the compound of interest, we could redistill the first fraction collected (from a clean flask!) and collect the first third of the material condensing in that process. This, the simplest of all fractional distillation strategies, is used in Example [2A]. High-performance fractional distillation is illustrated in Examples [3A] and [3B], pp. 83 and 87.

QUESTIONS

5-11. What is the major drawback of trying to distill a 500-µL mixture of liquids, all with boiling points below 200 °C?

5-12. How might you separate the mixture discussed in Question 5-6 if distillation were unsuccessful? Explain your choice.

5-13. A nonvolatile solute lowers the vapor pressure of the solvent in which it is dissolved. Why?

5-14. Why do simple distillations require that the components of the mixture to be separated have boiling points that are separated by 40 °C or more?

5-15. Which constituent of an equimolar mixture makes the larger contribution to the vapor pressure of the mixture, the higher or lower boiling component? Explain.

EXAMPLE [2A]

Simple Semimicroscale Distillation: Separation of Hexane and Toluene

Purpose

This experiment effects the separation of a binary liquid mixture composed of liquids having boiling points that are relatively far apart, greater than 30 °C. It will help you develop the skills to operate a semimicrodistillation apparatus so that purifications required in later experiments can be successfully carried out.

DISCUSSION

Hexane and toluene are liquid hydrocarbons that have boiling points approximately 40 °C apart. The liquid–vapor composition curve in Figure 5.6 represents this system; it is apparent that a two-plate distillation should yield nearly pure components. The procedure to be outlined consists of two parts. The first deals with the initial distillation (first plate), which separates the liquid mixture into three separate fractions. The second deals with *redistillation* of the first and third fractions (second plate). Exercising careful technique during the first distillation should provide a fraction rich in the lower boiling component, a middle fraction, and a fraction rich in the higher boiling component. Then careful *redistillation* of these fractions can be expected to complete the separation of the two components and to produce fractions of relatively pure hexane and toluene. The Hickman still used in the microscale laboratory is a simple, short-path column, and, therefore, one would not expect complete separation of the hexane and toluene in one cycle.

COMPONENTS

$$CH_3CH_2CH_2CH_2CH_2CH_3$$

Hexane

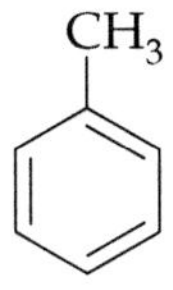

Toluene

EXPERIMENTAL PROCEDURE

Estimated time for the experiment: 2.0 h.

Physical Properties of Components

COMPOUND	MW	AMOUNT	bp (°C)	d	n_D
Hexane	86.18	1.0 mL	69	0.66	1.3751
Toluene	92.15	1.0 mL	111	0.87	1.4961

Reagents and equipment

Use an automatic delivery pipet to place 1.0 mL of hexane and 1.0 mL of toluene in a clean, dry, stoppered 5-mL conical vial.

Place the vial in a small beaker to prevent tipping. Add a boiling stone, assemble the Hickman still with the thermometer positioned directly down the center of the column (see previous discussion) and mount the system in a sand bath (see the equipment diagram, p. 78).

Experimental conditions

The temperature of the sand bath is raised to 80–90 °C, at a maximum rate of 5 °C/min (> 70 °C at 3 °C/min), using a hot plate.

Caution

> Do not let the temperature of the still rise too rapidly.

Once gentle boiling begins, the heating rate should be lowered to a maximum of 2 °C/min. It is *absolutely crucial* that the distillation rate be kept below 100 μL/3 min to achieve the necessary fraction enrichment that will permit good separation during the second stage of the experiment. The distillate is collected in *three fractions* over the temperature ranges (1) 65–85 °C (bath temperature ~95–110 °C); (2) 85–105 °C (bath temperature ~140 °C),) and (3) 105–110 °C (bath temperature ~170 °C) in amounts of approximately 800, 400, and 800 μL, respectively. Remove each fraction from the still with a bent-tip Pasteur pipet. Store the liquid condensate (fractions) in clean, dry, 1-dram, screw-cap vials. *Remember to number the vials in order and use an aluminum foil cap liner.*

Characterization of crude fractions

For each of the three fractions, record the refractive index. Fraction 1 has been enriched in one of the two components. Which one? Does the refractive index agree with that found in the literature? Fraction 3 has been enriched in the other

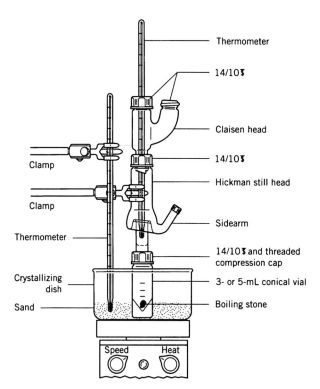

Hickman still with Claisen head adapter.

component. Does the refractive index of that fraction support your first conclusion? If partial enrichment has been achieved, proceed to the second phase of the distillation.

Redistillation of Fraction 1 Redistill fraction 1 in a clean Hickman still with a thermometer arranged as before (see equipment diagram), using a 3-mL conical vial and the procedure just outlined. Collect an initial fraction over the boiling range 68–71 °C (~100–200 μL). Remove it from the collar, using the Pasteur pipet, and place it in a 1-dram screw-cap vial.

Characterization of Fraction 1 Determine the ultramicro-boiling point and the refractive index of this lower boiling fraction. Compare the experimental values obtained with those of pure hexane reported in the literature.

Redistillation of Fraction 3 Fraction 3 is placed in a clean Hickman still, using a thermometer and a 3-mL conical vial, and redistilled using the procedure outlined. Collect an initial fraction over the boiling range 95–108 °C (~500 μL), and transfer this fraction by Pasteur pipet to a screw-cap vial. Collect a final fraction at 108–110 °C (~250 μL), and transfer the material to a second vial. *This second fraction is the highest boiling fraction to be collected in the three distillations and should be the richest in the high-boiling component.*

Characterization of Fraction 3 Determine the refractive index and boiling point of the second fraction and compare your results with those found in the literature for toluene. *Determine the refractive index and boiling point of pure toluene for comparison purposes.*

TECHNIQUE 3
Fractional Semimicroscale Distillation

Fractional distillation can occur in a distillation system containing more than one theoretical plate. This process must be used when the boiling points of the components differ by less than 30–40 °C and fairly complete separation is desired. A fractionating column is needed to accomplish this separation. As discussed previously, a liquid–vapor composition curve (Fig. 5.8) shows that the

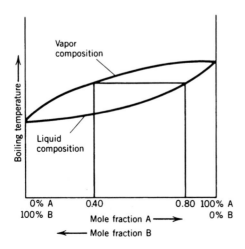

Figure 5.8 Liquid–vapor composition curve.

lower boiling component of a binary mixture makes a larger contribution to the vapor composition than does the higher boiling component. On condensation, the liquid formed will be richer in the lower boiling component. This condensate will not be pure, however, and, in the case of components with close boiling points, it may be only slightly enriched. If the condensate is vaporized a second time, the vapor in equilibrium with this liquid will show a further enrichment in the lower boiling component. The trick to separating liquids with similar boiling points is to repeat the vaporization–condensation cycle many times. Each cycle is one *theoretical plate*. Several column designs, which achieve varying numbers of theoretical plates, are available for use at the macro level (Fig. 5.9).

Most distillation columns are designed so that fractionation efficiency is achieved by the very large surface area in contact with the vapor phase (and very similar to the way increased resolution is obtained on a GC column (see Technique 1). This increased surface area can be accomplished by packing the fractionating column with wire gauze or glass beads. Unfortunately, a large volume of liquid must be distributed over the column surface in equilibrium with the vapor. Furthermore, the longer the column the more efficient it becomes (see Technique 1), but longer columns also require additional liquid phase. The amount of liquid phase required to fill the column with a liquid–vapor equilibrium is called *column holdup*. Column holdup is essentially lost from the distillation because this volume can never go past the top

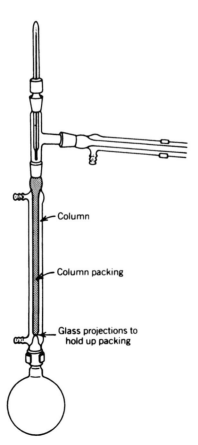

Column

Column packing

Glass projections to
hold up packing

Figure 5.9 A fractional distillation setup. *(From Zubrick, J. W. The Organic Chem Lab Survival Manual, 4th ed.; Wiley: New York, 1997. Reprinted with permission of John Wiley & Sons, Inc., New York.)*

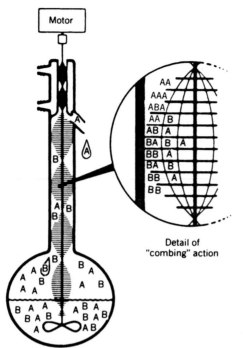

Figure 5.10 Schematic of a metal-mesh, spinning-band still.

of the distillation column; it can only return to the distillation pot upon cooling. Column holdup can be large compared to the total volume of material available for the distillation. With mixtures of less than 2 mL, column holdup precludes the use of the most common fractionation columns. Columns with rapidly spinning bands of metal gauze or Teflon have very low column holdup and have a large number of theoretical plates relative to their height (Fig. 5.10).

Microscale spinning-band distillation apparatus (Fig. 5.11) can achieve nearly 12 theoretical plates and are simple enough to be used in the instructional laboratory. This still has a Teflon band that fits closely inside an insulated glass tube. The Teflon band has spiral grooves which, when the band is spun (1000–1500 rpm), rapidly return condensed vapor to the distillation pot. A powerful extension of this apparatus uses a short spinning band inside a modified Hickman still head (see Fig. 3.15, p. 28). These stills are called Hickman–Hinkle stills; 4-cm Hickman–Hinkle columns can have more than 10 theoretical plates. The commercially available 2.5-cm version is rated at six theoretical plates. Examples [3A] and [3B] involve fractional distillations with spinning-band columns.

The thermometer is positioned directly down the center of the distillation column, with the bulb just at the bottom of the well. It is very important to position both the still and the thermometer as vertically as possible; the thermometer must not touch the glass walls of the column (Fig. 5.12). Example [2A] uses the Hickman still for a set of simple distillations that overall yield a fractional distillation. A two-theoretical-plate distillation can in effect be obtained with this system on a two-component mixture by carrying out two sequential fractional distillations.

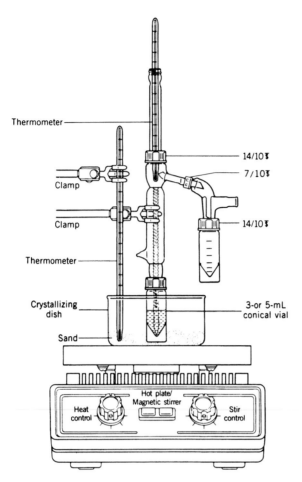

Thermometer

Clamp

Clamp

Thermometer

Crystallizing dish

Sand

14/10 ꟊ

7/10 ꟊ

14/10 ꟊ

3-or 5-mL conical vial

Heat control

Hot plate/ Magnetic stirrer

Stir control

Figure 5.11 Microspinning band distillation column (3 in.).

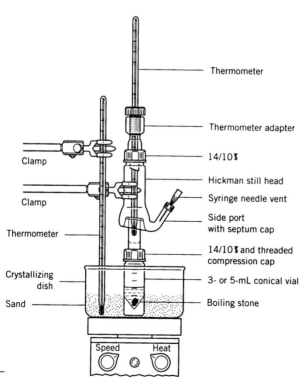

Thermometer

Clamp

Clamp

Thermometer

Crystallizing dish

Sand

Thermometer adapter

14/10 ꟊ

Hickman still head

Syringe needle vent

Side port with septum cap

14/10 ꟊ and threaded compression cap

3- or 5-mL conical vial

Boiling stone

Speed   Heat

Figure 5.12 Hickman still with thermometer adapter.

For a more detailed discussion of how spinning bands work, see the discussion of distillation, 🌀.

QUESTIONS

5-16. Why is it very important that the hot vapor in microscale distillations climb the column very slowly?

5-17. Why might Teflon be the material of choice for constructing microscale spinning bands?

5-18. The spinning band overcomes two major problems of microscale distillations by wiping the liquid condensate rapidly from the column walls. What are these problems?

5-19. Why are spinning bands so effective at increasing the number of theoretical plates in distillation columns?

5-20. Why is steam distillation often used to isolate and purify naturally occurring plant substances?

EXAMPLE [3A]

Fractional Semimicroscale Distillation: Separation of 2-Methylpentane and Cyclohexane Using a Spinning-Band Column

Purpose

The purpose of this experiment is to separate two liquids with boiling points that are relatively similar: less than 20 °C apart, to learn the operation of a high-performance spinning-band distillation column, and to develop the skills for purifying small quantities of liquid mixtures.

DISCUSSION

In this experiment, the separation of a 2-mL mixture of 2-methylpentane and cyclohexane using a 2.5-in. spinning-band distillation column is described. The purity of the fractions is determined by gas chromatography and by measurement of the refractive index. Finally, the number of theoretical plates is estimated. You will separate a 50:50 mixture of 2-methylpentane and cyclohexane using a spinning band distillation column shown on p. 84.

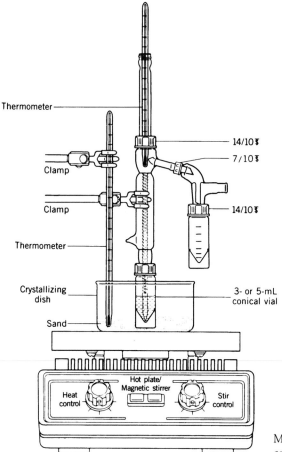

Micro spinning-band distillation column (2.5 in.).

COMPONENTS

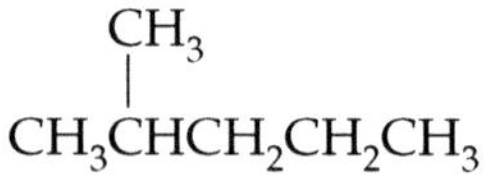

CH$_3$
CH$_3$CHCH$_2$CH$_2$CH$_3$

2-Methylpentane

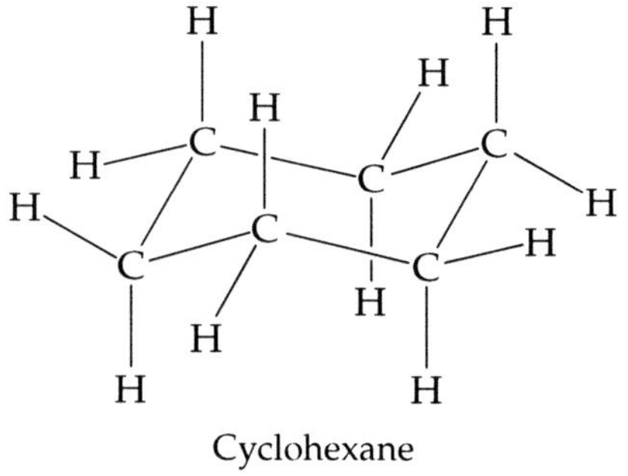

Cyclohexane

EXPERIMENTAL PROCEDURE

Estimated time for the experiment: 3.0 h.

Physical Properties of Components

COMPOUND	MW	AMOUNT	bp (°C)	n_D
2-Methylpentane	86.18	1.0 mL	60.3	1.3715
Cyclohexane	84.16	1.0 mL	80.7	1.4266

Reagents and equipment

Assemble the system as shown in Figure 3.16 making sure that the Teflon band is aligned as straight as possible in the column. In particular, the pointed section extending into the vial must be straightened to minimize vibration during spinning of the band.

Place a pipet bulb on the side arm of the collection adapter. This bulb plays an important function in the operation of the column: Attachment of the bulb creates a closed system. Suspension of the thermometer with a septum on the top of the condenser can act to release any buildup of pressure.

Caution

The system must be able to vent at the thermometer during operation!

Once the spinning band has been tested and rotates freely, place 1.0 mL of 2-methylpentane and 1.0 mL of cyclohexane in the vial (to be delivered with a Pasteur pipet or an automatic delivery pipet). Reassemble the system and lower the column into the sand bath or copper-tube block. The beveled edge on the air condenser should be rotated 180° from the collection arm.

Note

It is important to make an aluminum foil cover for the sand bath; this cover will reflect the heat and hot air away from the collection vial.

Experimental conditions

Gently heat (copper-tube block, ⬤ Fig. 3.3W) the vial until boiling occurs. The magnetic stirrer is turned to a low-spin rate when heating commences. When reflux is observed at the base of the column, the magnetic stirrer is adjusted to intermediate spin rate. Once liquid begins to enter the column the spin rate is increased to the maximum (1000–1500 rpm).

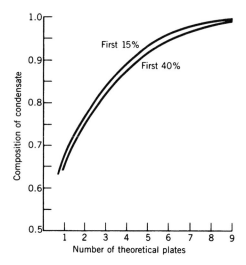

Composition of the first 15% and the first 40% of the volume collected in the distillation of a 50% (v/v) mixture of 2-methylpentane and cyclohexane.

Note

It is absolutely critical that the temperature of the vial be adjusted so that vapors in the column rise very slowly. It is possible for over-heated vapors to be forced through the air condenser.

When the vapors slowly arrive in the unjacketed section of the column head, the condenser joint acts as a vapor shroud to effectively remove vapors from the receiver-cup area. During this total reflux period, maximum separation of the components is achieved. Once vapor reflux occurs in the head of the column, the system is left for 20–30 min to reach thermal equilibrium. During this period of total reflux, the head thermometer should read about 57–60 °C (at least for most of the equilibration time).

Following the equilibration period, collection of the resolved components may begin. Rotate the air condenser 180° so that the beveled edge is over the collection duct. At this point, manipulation of the pipet bulb allows drainage of the condensate from the side arm. (This procedure is repeated occasionally to continue drainage from the side arm.) Collect six drops (~0.30 mL). After removing the collection vial, transfer the contents into a covered vial using a Pasteur pipet. Label all fractions. Collect two 0.6-mL fractions (the pipet bulb may be removed during collection of these latter fractions); then turn off the heat and stirring motor, and remove the vial from the sand bath. Transfer the material remaining in the vial, using a Pasteur pipet, to a fourth covered vial.

Characterization of the fractions

The composition of each of the fractions may be determined by gas chromatographic analysis and measurement of the refractive index. A GOW-MAC gas chromatograph should be set up as follows:

Column	DC 710
Injection	10 μL
Temperature	80 °C
Flow rate (He)	55 mL/min
Chart speed	1 cm/min

If we assume that the refractive index is a linear function of the volume fraction, the following relationship gives us the volume fraction of 2-methylpentane in a mixture. The volume fraction is X and the measured refractive index is n_D:

$$X = \frac{1.4266 - n_D}{1.4266 - 1.3715}$$

The curve shown in the graph of composition of condensate vs. number of theoretical plates (above) may be used to estimate the number of theoretical plates from the composition (mole fraction) of the *first* 0.30-mL fraction. For example, if the composition of the *first* 0.30 mL is 0.89, we would infer that the system had a resolution equivalent to about four theoretical plates. Note that the number of plates cannot be determined with confidence if the composition is greater than about 0.97. If we really wanted to determine the number of theoretical plates for a system with more than five plates, we could start with a mixture containing only 10 or 20% of the most volatile component (MVC), rather than the 50% used in this example.

EXAMPLE [3B]

Fractional Semimicroscale Distillation: The Separation of 2-Methylpentane and Cyclohexane Using a Spinning Band in a Hickman–Hinkle Still

Purpose

In this exercise you will become familiar with a powerful modification of the classic Hickman still: the Hickman–Hinkle spinning band distillation apparatus. This small still is one of the most efficient techniques developed for the purification of small quantities of liquids. You will develop the skills for handling small quantities of liquids and their purification by distillation, and become familiar with these techniques, so that they may be used in the purification of reaction products formed in later experiments.

DISCUSSION

This distillation separates the same two compounds used in Example [2B], p. 83. The distillate can be analyzed to determine the number of theoretical plates. If careful attention is given to the procedure, the spinning Hickman–Hinkle is capable of more than six theoretical plates.

As in Example [3A] the separation of a 2-mL mixture of 2-methylpentane and cyclohexane is achieved. The purity of the fractions can be determined by gas chromatography and by measurement of the refractive index.

COMPONENTS

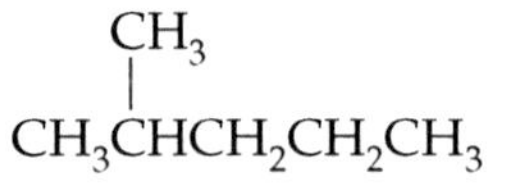

CH$_3$
|
CH$_3$CHCH$_2$CH$_2$CH$_3$

2-Methylpentane 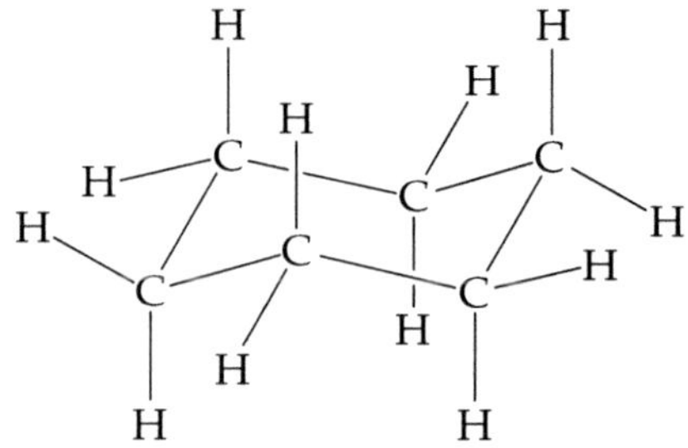   Cyclohexane

EXPERIMENTAL PROCEDURE

Estimated time for the experiment: 3.0 h.

Physical Properties of Components

COMPOUND	MW	AMOUNT	bp (°C)	n_D
2-Methylpentane	86.18	1.0 mL	60.3	1.3715
Cyclohexane	84.16	1.0 mL	80.7	1.4266

Reagents and equipment

The system is assembled as shown in Figure 3.15. In the process, make sure that the Teflon band is aligned as straight as possible in the column. In particular, the pointed section extending into the vial must be straightened to minimize vibration during spinning of the band.

Once the spinning band has been tested and rotates freely, place 1.0 mL of 2-methylpentane and 1.0 mL of cyclohexane in the vial (to be delivered with a Pasteur pipet or an automatic delivery pipet). Reassemble the system and lower the column into the sand bath.

Cover the sand bath with aluminum foil during the distillation to prevent the collar of the still from overheating. It is easier to regulate the temperature of the bath when it is covered. However, for distillations, a more efficient heating technique is the recently developed copper-tube block (🌀, Fig. 3.3W).

Experimental conditions

Gently heat the vial until boiling occurs. When heating commences, turn on the magnetic stirrer at a low setting. When reflux commences at the base of the column the magnetic stirrer is raised to intermediate settings. Once liquid begins to enter the column, the spin rate is increased to the maximum (1000–1500 rpm).

Note

> It is extremely important that careful temperature control be exercised at this stage so that the condensing vapors ascend the column very slowly.

Vapor-phase enrichment by the most volatile component is limited mainly to this period, as fraction collection commences immediately on arrival of the vapor column at the annular ring. Once condensation occurs, fractions are collected by the same technique used in Example [2A], p. 76. Characterization of the fractions, however, follows the procedure given in Example [3A], p. 83.

Characterization of the fractions

The composition of each of the fractions may be determined by gas chromatography, the refractive index, or both. See Example [3A] for details.

An alternative approach to the procedures discussed in Example [3B] is to establish the fraction volume by weight. The curves shown on p. 86 may again be used to estimate the number of theoretical plates. The volume of the first fraction can be estimated, or determined more accurately by weighing the fraction in a tared screw-cap vial. The composition of this fraction then may be determined and the fraction of the total represented by this portion calculated. If, for example, the first fraction has a volume of 0.4 mL (20% of the total) and has a composition 0.89 by volume of 2-methylpentane, we would infer that the system had a resolution equivalent to about four theoretical plates.

QUESTIONS

5-21. The boiling point of a liquid is affected by several factors. What effect does each of the following conditions have on the boiling point of a given liquid?
   (a) The pressure of the atmosphere
   (b) Use of an uncalibrated thermometer
   (c) Rate of heating of the liquid in a distillation flask

5-22. Calculate the vapor pressure of a solution containing 30 mol% hexane and 70 mol% octane at 90 °C, assuming that Raoult's law is obeyed.
   Given: vapor pressure of the pure compounds at 90 °C: hexane = 1390 torr, octane = 253 torr.

5-23. In any distillation for maximum efficiency of the column, the distilling flask should be approximately one-half full of liquid. Comment on this fact in terms of (a) a flask that is too full and (b) a flask that is nearly empty.

5-24. Occasionally during a distillation, a solution will foam rather than boil. A way of avoiding this problem is to add a *surfactant* to the solution.
   (a) What is a surfactant?
   (b) What is the chemical constitution of a surfactant?
   (c) How does a surfactant reduce the foaming problem?

5-25. Explain why packed and spinning-band fractional distillation columns are more efficient at separating two liquids with close boiling points than are unpacked columns.

5-26. Explain what effect each of the following mistakes would have had on the simple distillation carried out in this experiment.

(a) You did not add a boiling stone.

(b) You heated the distillation flask at too rapid a rate.

5-27. In the ultramicro-boiling-point determination, why is the boiling point taken just as bubbles cease emerging from the bell?

5-28. Define each of the following terms, which are related to the distillation process:

(a) Distillate

(b) Normal boiling point

(c) Forerun

5-29. How does the refractive index of a liquid vary with temperature? What corrective factor is often used to determine the value at a specific temperature, for example, 20 °C, if the measurement were made at 25 °C?

BIBLIOGRAPHY

General references on distillation

LODWIG, S. N. *J. Chem. Educ.* **1989**, *66*, 77.

STICHLMAIR, J.; FAIR, J. R. *Distillation: Principles and Practice*, Wiley: New York, 1998.

Technique of Organic Chemistry, Vol. IV, Distillation, 2nd ed., PERRY, E. S.; WEISSBERGER, A.; Interscience-Wiley, New York, 1967.

VOGEL, A. I. *Vogel's Textbook of Practical Organic Chemistry*, 5th ed.; FURNIS, B. S., et al. Eds.; Wiley: New York, 1989.

ZUBRICK, J. W. *The Organic Chem Lab Survival Manual*, 4th ed.; Wiley: New York, 1997.

TECHNIQUE 4

Solvent Extraction

Solvent extraction is frequently used in the organic laboratory to separate or isolate a desired compound from a mixture or from impurities. Solvent extraction methods are readily adapted to microscale work because small quantities are easily manipulated in solution. Solvent extraction methods are based on the solubility characteristics of organic substances in the solvents used in a particular separation procedure. Liquid–liquid and solid–liquid extractions are the two major types of extractions used in the organic laboratory.

Solubility

Substances vary greatly in their solubility in various solvents, but a useful and generally true principle is that a substance tends to dissolve in a solvent that is chemically similar to itself. In other words, *like dissolves like*. The significant exceptions to this general statement are seen when solubilities are determined by acid–base properties.

Thus, to be soluble in water a compound needs to have some of the molecular characteristics of water. Alcohols, for example, have a hydroxyl group (—OH) bonded to a hydrocarbon chain or framework (R—OH). The hydroxyl group can be thought of as effectively half a water (H_2O) molecule; its polarity is similar to that of water. This polarity is due to the charge separation arising from the different electronegativities of the hydrogen and oxygen atoms. The O—H bond, therefore, is considered to have partial ionic character.

$$\overset{\delta^-}{\underset{..}{\overset{..}{O}}}\!\!-\!\!\overset{\delta^+}{H}$$

Partial ionic character of the hydroxyl group

This polar, or partial ionic, character leads to relatively strong hydrogen bond formation between molecules with hydroxyl groups. Strong hydrogen bonding (shown here for the ethanol–water system) occurs in molecules that have a hydrogen atom attached to an oxygen, nitrogen, or fluorine atom—all three are quite electronegative atoms.

Ethanol   Hydrogen bond formation

The hydroxyl end of the ethanol molecule is very similar to water. When ethanol is added to water, therefore, they are miscible in all proportions. That is, ethanol is completely soluble in water and water is completely soluble in ethanol. This degree of solubility occurs because the attractive forces between the two molecules are nearly as strong as those between two water molecules; however, the attraction in the first case is somewhat weakened by the presence of the nonpolar ethyl group,—CH_2CH_3. Hydrocarbon groups attract each other only weakly, as demonstrated by their low melting and boiling points. Three examples of the contrast in boiling points between compounds of different structure, but similar molecular weight, are summarized in Table 5.1. Molecules that attract each other weakly (lower intermolecular forces) have lower boiling points.

Ethanol is completely miscible with water, but the solubility of octanol in water is less than 1%. Why the difference in solubilities between these two alcohols? The dominant structural feature of ethanol is its polar hydroxyl group; the dominant structural feature of octanol is its nonpolar alkyl group:

Octanol

$$CH_3\!\!-\!\!CH_2\!\!-\!\!\overset{..}{\underset{..}{O}}\!\!-\!\!CH_2\!\!-\!\!CH_3$$

Diethyl ether

As the size of the hydrocarbon section of the alcohol molecule increases, the intermolecular attraction between the polar hydroxyl groups of the alcohol and the water molecules is no longer strong enough to overcome the hydrophobic (lacking attraction to H_2O) nature of the nonpolar hydrocarbon section of the alcohol. On the other hand, octanol has a large nonpolar hydrocarbon group as its dominant structural feature. We might, therefore, expect octanol to be more soluble in less polar solvents, and, in fact, octanol is completely miscible with diethyl ether. Ethers are weakly polar solvents because a

TABLE 5.1 *Comparison of Boiling Point Data*

NAME	FORMULA	MW	bp (°C)
Ethanol	CH_3CH_2OH	46	78.3
Propane	$CH_3CH_2CH_3$	44	−42.2
Methyl acetate	$CH_3CO_2CH_3$	74	54
Diethyl ether	$(CH_3CH_2)_2O$	74	34.6
Ethene	$CH_2=CH_2$	28	−102
Methylamine	CH_3NH_2	31	−6

C—O bond is much less polar than an O—H bond (carbon is less electronegative than oxygen). Because both octanol and diethyl ether are rather nonpolar, each is completely soluble in the other. For compounds with both polar and nonpolar groups, in general, those compounds with five or more carbon atoms in the hydrocarbon portion of the molecule will be more soluble in nonpolar solvents, such as pentane, diethyl ether, or methylene chloride. Figure 5.13 summarizes the solubilities of a number of straight-chain alcohols, carboxylic acids, and hydrocarbons in water. As expected, most monofunctional compounds with more than five carbon atoms have solubilities similar to the hydrocarbons.

Several additional relationships between solubility and structure have been observed.

1. Branched-chain compounds have greater water solubility than their straight-chain counterparts, as illustrated in Table 5.2 with a series of alcohols.

2. The presence of more than one polar group in a compound will increase that compound's solubility in water and decrease its solubility in nonpolar solvents. For example, sugars, such as cellobiose, contain multiple hydroxyl and/or acetal groups and are water soluble and ether insoluble. Cholesterol, which has only a single hydroxyl group on its 27 carbon atoms, is insoluble in water and quite soluble in ether.

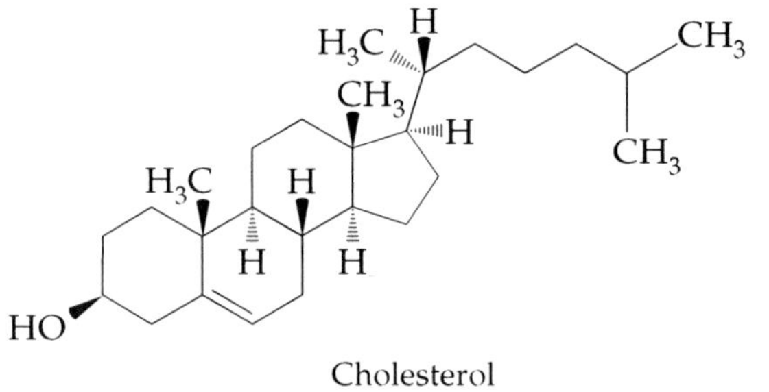

Cholesterol

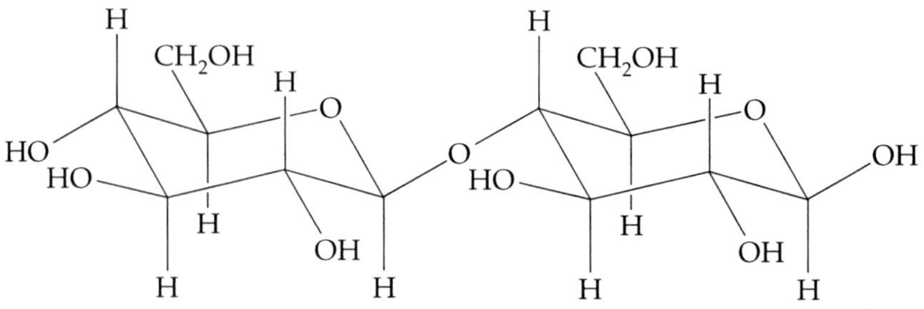

Cellobiose

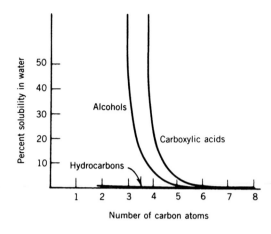

Figure 5.13 Solubility curve of acids, alcohols, and hydrocarbons. *(From Kamm, O.* Qualitative Organic Analysis, *2nd ed,; Wiley: New York, 1932. Reprinted with premission of John Wiley & Sons, New York.)*

TABLE 5.2 *Water Solubility of Alcohols*

NAME	STRUCTURAL FORMULA	SOLUBILITY (g/100 g H_2O at 20 °C)
Hexanol	$CH_3(CH_2)_4CH_2OH$	0.6
Pentanol	$CH_3(CH_2)_3CH_2OH$	2.2
2-Pentanol	$CH_3(CH_2)_2CH(OH)CH_3$	4.3
2-Methyl-2-butanol	$(CH_3)_2C(OH)CH_2CH_3$	11.0

3. The presence of a chlorine atom, even though it lends some partial ionic character to the mostly covalent C—Cl bond, does not normally impart water solubility to a compound. In fact, compounds such as methylene chloride (CH_2Cl_2), chloroform ($CHCl_3$), and carbon tetrachloride (CCl_4) have long been used as solvents for extracting aqueous solutions. The latter two solvents are not often used nowadays, unless strict safety precautions are exercised, because they are potentially carcinogenic.

4. Most functional groups capable of forming a hydrogen bond with water increased the water solubility of a substance. For example, smaller alkyl amines have significant water solubility; the water-solubility data for a series of amines are summarized in Table 5.3.

The solubility characteristics of any given compound govern its distribution (*partition*) between the phases of two immiscible solvents (in which the material has been dissolved) when these phases are intimately mixed.

TABLE 5.2 *Water Solubility of Amines*

NAME	STRUCTURAL FORMULA	SOLUBILITY (g/100 g H_2O at 25 °C)
Ethylamine	$CH_3CH_2NH_2$	∞
Diethylamine	$(CH_3CH_2)_2NH$	∞
Trimethylamine	$(CH_3)_3N$	91
Triethylamine	$(CH_3CH_2)_3N$	14
Aniline	$C_6H_5—NH_2$	3.7
1,4-Diaminobenzene	$H_2N—C_6H_4—NH_2$	3.8

PARTITION (OR DISTRIBUTION) COEFFICIENT

A given substance X is partially soluble in each of two immiscible solvents. If X is placed in a mixture of these two solvents and shaken, an equilibrium will be established between the two phases. That is, substance X will partition (distribute) itself in a manner that is a function of its relative solubility in the two solvents:

$$X_{\text{solvent 1}} \rightleftharpoons X_{\text{solvent 2}}$$

The equilibrium constant, K_p, for this equilibrium expression is known as the *partition* or *distribution coefficient*:

$$K_p = \frac{[X_{\text{solvent 2}}]}{[X_{\text{solvent 1}}]}$$

The equilibrium constant is thus the ratio of the concentrations of the species, X, in each solvent for a given system at a given temperature. The partition coefficient can be conveniently estimated as the ratio of the solubility of X in solvent 1 vs. solvent 2:

$$K_p = \frac{\text{solubility of X in solvent 2}}{\text{solubility of X in solvent 1}}$$

When solvent 1 is water and solvent 2 is an organic solvent such as diethyl ether, the basic equation used to express the coefficient K_p is

$$K_p = \frac{(\text{g}/100 \text{ mL})_{\text{organic layer}}}{(\text{g}/100 \text{ mL})_{\text{water layer}}}$$

This expression uses grams per 100 mL for the concentration units. Note that the partition coefficient is dimensionless, so any concentration units may be used if the units are the same for both phases. For example, grams per liter (g/L), parts per million (ppm), and molarity (M) can all be used. If equal volumes of both solvents are used, the equation reduces to the ratio of the weights ($g_{\text{organic}}/g_{\text{water}}$) of the given species in the two solvents:

$$K_p = \frac{g_{\text{organic layer}}}{g_{\text{water layer}}}$$

Determination of the partition coefficient for a particular compound in various immiscible-solvent combinations often can give valuable information for isolating and purifying the compound by using extraction techniques. Thus, liquid–liquid extraction is a common separation technique used in organic as well as analytical laboratories.

Table 5.4 provides some examples of K_p values determined at room temperature for a number of compounds in the water–methylene chloride system.

Let us now look at a typical calculation for the extraction of an organic compound P from an aqueous solution using diethyl ether. We will assume that the $K_{p \text{ ether/water}}$ value (partition coefficient of P between diethyl ether and water) is 3.5 at 20 °C. If a solution of 100 mg of P in 300 μL of water is extracted at 20 °C with 300 μL of diethyl ether, the following expression holds:

TABLE 5.4 *Representative K_p Values in CH_2Cl_2—H_2O*

COMPOUND	K_P VALUE
Nitrobenzene	51.5
Aniline	3.3
1,2-Dihydroxybenzene	0.2

$$K_{\text{p ether/water}} = \frac{C_e}{C_w} = \frac{W_e/300\ \mu L}{W_w/300\ \mu L}$$

where W_e = weight of P in the ether layer

W_w = weight of P in the water layer

C_e = concentration of P in the ether layer

C_w = concentration of P in the water layer

Since $W_w = 100 - W_e$, the preceding relationship can be written as

$$K_{\text{p ether/water}} = \frac{W_e/300\ \mu L}{(100 - W_e)/300\ \mu L} = 3.5$$

If we solve for the value of W_e, we obtain 77.8 mg; the value for W_w is 22.2 mg. Thus, we see that after one extraction with 300 μL of ether, 77.8 mg of P (77.8% of the total) is removed by the ether and 22.2 mg (22.2% of the total) remains in the water layer. Is it preferable to make a single extraction with the total quantity of solvent available, or to make multiple extractions with portions of the solvent? The second method is usually more efficient. To illustrate, consider extracting the 100 mg of P in 300 μL of water with *two* 150-μL portions of diethyl ether instead of one 300-μL portion.

For the first 150-μL extraction,

$$\frac{W_e/150\ \mu L}{W_w/300\ \mu L} = \frac{W_e/150\ \mu L}{(100 - W_e)/300\ \mu L}$$

Solving for W_e, we obtain 63.6 mg. The amount of P remaining in the water layer (W_w) is then 36.4 mg. The aqueous solution is now extracted with the second portion of ether (150 μL). We then have

$$\frac{W_e/150\ \mu L}{(36.4 - W_e)/300\ \mu L} = 3.5$$

As before, by solving for W_e, we obtain 23.2 mg for the amount of P in the ether layer; $W_w = 13.2$ mg in the water layer.

The two extractions, each with 150 μL of ether, removed a total of 63.6 mg + 23.2 mg = 86.8 mg of P (86.8% of the total). The P left in the water layer is then 100 − 86.8, or 13.2 mg (13.2% of the total).

It can be seen from these calculations that the multiple-extraction technique is more efficient. The single extraction removed 77.8% of P; the double extraction (with the same total volume of ether) increased this to 86.8%. To extend this relationship, three extractions with one-third the total quantity of ether in each portion would be even more efficient. You might wish to calculate this to prove the point. Of course, there is a practical limit to the number of extractions that can be performed.

The multiple-extraction example shown here illustrates that several extractions with small volumes is more efficient than a single extraction procedure. This is always true *provided* the partition coefficient is neither very large nor very small. If the partition coefficient K_p for a substance between two solvents is very large ($K_p > 100$) or very small ($K_p < 0.01$) multiple extractions (using the same total amount of solvent) do not significantly increase the efficiency of the extraction process.[5]

Extraction

LIQUID–LIQUID EXTRACTION

The more common type of extraction, liquid–liquid extraction, is used extensively. It is a very powerful method for separating and isolating materials at the microscale level. It is operationally not a simple process, so attention to detail is critical.

There are several important criteria to consider when choosing a solvent for the extraction and isolation of a component from a solution:

* The chosen extraction solvent must be immiscible with the solution solvent.
* The chosen extraction solvent must be favored by the distribution coefficient for the component being extracted.
* The chosen extraction solvent should be readily separated from the desired component after extraction. This usually means that it should have a low boiling point.
* The chosen organic extraction solvent must not react chemically with any component in the aqueous mixture being extracted.

Note

The aqueous phase may be modified, as in acid–base extractions, but the organic solvent does not react with the components in the aqueous mixture.

MICROSCALE EXTRACTION

A capped conical vial or a stoppered centrifuge tube is the best container for most microscale extractions, but a small test tube may be used. Note that a conical vial and a centrifuge tube have the same inner shape. This shape has the advantage that as the lower phase (layer) is withdrawn by pipet, the interface (boundary) between the two liquid phases becomes narrower and narrower, and thus easier to see, at the bottom of a conical container (see Figure 5.15). This is not the case for a test tube. The centrifuge tube has the added advantage that if a solid precipitate must be separated or an emulsion broken up, it can easily be done using a centrifuge.

---

[5]Palleros, D. R. *J. Chem. Educ.* **1995**, *72*, 319.

A good rule of thumb is that the container to be used for the extraction should be *at least three times the volume of liquid* you wish to extract.

Regardless of the container used, in any liquid–liquid extraction, the two immiscible solvents must be completely mixed to maximize the surface area of the interface between the two and allow partitioning of the solute. This can be accomplished by shaking (carefully to avoid leakage around the cap), using a Vortex mixer, or by adding a magnetic spin vane and then stirring with a magnetic stirrer.

Another important rule in the extraction process is that you should *never discard any layer until the isolation is complete.*

Let us consider a practical example. Benzanilide can be prepared by the in situ rearrangement of benzophenone oxime in acid solution:

Benzophenone   Benzophenone oxime   Benzanilide

The benzanilide is separated from the reaction mixture by extraction with three 1.0-mL portions of methylene chloride solvent.

Note

Saying, for example, "extracted with three 1.0-mL portions of methylene chloride" means that three extractions are performed (one after the other, each using 1.0 mL of methylene chloride) and the three methylene chloride extracts are combined.

A microscale extraction process consists of two parts: (1) mixing the two immiscible solutions, and (2) separating the two layers after the mixing process.

1. *Mixing.* In the experimental procedure for the isolation of the benzanilide product, methylene chloride (1.0 mL) is added to the aqueous reaction mixture contained in a 5.0-mL conical vial (or centrifuge tube). The extraction procedure is outlined in the following steps:

Step 1. Cap the vial.

Step 2. Shake the vial gently to thoroughly mix the two phases (careful!)

Note

The mixing may be carried out using a Vortex mixer or magnetic stirrer—see previous discussion.

Step 3. Carefully vent the vial by loosening the cap to release any pressure that may have developed.

Step 4. Allow the vial to stand on a level surface to permit the two phases to separate. A sharp phase interface should appear.

Note _____

For safety reasons it is advisable to place the vial in a small beaker to prevent tipping. If a volatile solvent such as ether is used, it is advisable to place the vial or centrifuge tube in a beaker of ice water to prevent loss of solvent during the transfers.

2. *Separation.* At the microscale level, the two phases are separated with a Pasteur filter pipet (a simple Pasteur pipet can be used in some situations), which acts as a miniature separatory funnel. The separation of the phases is shown in Figure 5.14.

A major difference between macro and micro techniques is that when microscale volumes are used, as just discussed, the mixing and separation are done in two parts. When macroscale volumes are used in a separatory funnel, mixing and separation are both done in the funnel in one step. The separatory funnel is an effective device for extractions with larger volumes, but it is not practical for microscale extractions because of the large surface areas involved. The recommended procedures are shown in Figures 5.15 and 5.16.

Benzanilide is more soluble in methylene chloride than in water. Multiple extractions are performed to ensure complete removal of the benzanilide from the aqueous phase. The methylene chloride solution is the lower layer because it is more dense than water. The following list outlines the general method for an organic solvent more dense than water (refer to Fig. 5.15).

Note _____

(a) A pipet pump may be used to replace the bulb. (b) One technique is to hold the pipet across the palm of the hand and squeeze the bulb with the thumb and index finger. (c) Remember to have an empty tared vial available in which to place the separated phase.

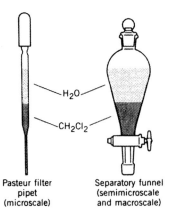

Pasteur filter
pipet
(microscale)

Separatory funnel
(semimicroscale
and macroscale)

Figure 5.14 Extraction devices.

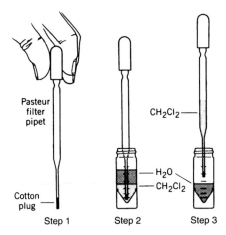

Figure 5.15 Pasteur filter pipet separation of two immiscible liquid phases; the more dense layer contains the product.

Step 1. Squeeze the pipet bulb to force air from the pipet.

Step 2. Insert the pipet into the vial until it is close to the bottom. Be sure to hold the pipet vertically.

Step 3. Carefully allow the bulb to expand, drawing only the lower methylene chloride layer into the pipet. This should be done in a smooth, steady manner so as not to disturb the interface between the layers. With practice, you can judge the amount that the bulb must be squeezed to just separate the layers. Keep the pipet vertical. *Do not tip the pipet back and allow liquid to enter the bulb! Do not suck liquid into the bulb!*

Step 4. (Step 4 is not shown in the figure.) Holding the pipet vertical, place it over and into the neck of an empty vial (as shown in Fig. 5.16, Step 2), and gently squeeze the bulb to transfer the methylene chloride solution into the vial. A second extraction can now be performed after adding another portion of methylene chloride to the original vial. The procedure is repeated. Multiple extractions can be performed in this manner. Each methylene chloride extract is transferred to the same vial—that is, the extracts are combined. The reaction product has now been transferred from the aqueous layer (aqueous phase) to the methylene chloride layer (organic phase), and the phases have been separated.

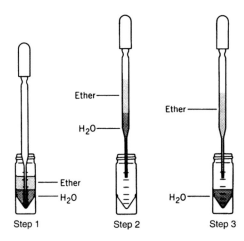

Figure 5.16 Pasteur filter pipet separation of two immiscible liquid phases; the less dense layer contains the product.

In a diethyl ether–water extraction, the ether layer is less dense and thus is the upper layer (phase). An organic reaction product generally dissolves in the ether layer and is thus separated from water-soluble byproducts and other impurities. The procedure followed to separate the water–ether phases is identical to that described above for methylene chloride–water systems, except that here the top layer (organic layer) is transferred to the new container. The following list outlines the general method for an organic solvent less dense than water (refer to Fig. 5.16).

Step 1. Squeeze the pipet bulb to force air from the pipet and insert the pipet into the vial until it is close to the bottom. Then, draw **both** phases slowly into the pipet. Keep the pipet vertical. *Do not tip the pipet back and allow liquid to enter the bulb! Do not suck liquid into the bulb!* Try not to allow air to be sucked into the pipet, as this tends to mix the phases in the pipet. If mixing does occur, allow time for the interface to re-form.

Step 2. Return the aqueous layer (bottom layer) to the **original** container by gently squeezing the pipet bulb.

Step 3. Transfer the separated ether layer (top layer) to a new tared vial.

SEPARATORY FUNNEL—SEMIMICROSCALE
AND MACROSCALE EXTRACTIONS

A separatory funnel (Fig. 5.14) is effective for extractions carried out at the semimicroscale and macroscale levels. The mixing and separation are done in the funnel itself in one step. Many of you may be familiar with this device from the general chemistry laboratory. The same precautions as outlined above for microscale extraction should be observed here.

Note

The funnel size should be such that the total volume of solution is less than half the total volume of the funnel. If the funnel has a ground-glass stopcock and/or stopper, the ground-glass surfaces must be lightly greased to prevent sticking, leaking, or freezing. If Teflon stoppers and stopcocks are used, grease is not necessary because these are self-lubricating.

Step 1. Close the stopcock of the separatory funnel.

Step 2. Add the solution to be extracted, after first making sure that the *stopcock is closed.* The funnel should be supported in an iron ring attached to a ring stand or rack on the lab bench.

Step 3. Add the proper amount of extraction solvent (about one-third of the volume of the solution to be extracted is a good rule of thumb) and place the stopper on the funnel.

Step 4. Remove the funnel from the ring stand, keeping the stopper in place with the index finger of one hand, and holding the funnel in the other hand with your fingers positioned so they can operate the stopcock (Fig. 5.17*a*).

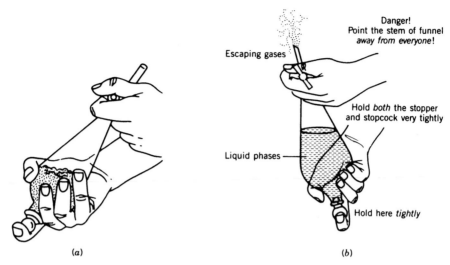

Figure 5.17 (a) Correct position for holding a separatory funnel while shaking. (b) Correct method for venting a separatory funnel.

Step 5. Carefully invert the funnel (make sure its stem is pointing up, and not pointing at you or anyone else). Slowly open the stopcock to release any built-up pressure (Fig. 5.17*b*). Close the stopcock and then shake the funnel for several seconds. Position the funnel for venting (make sure the stem is pointing up, and not pointing at you or anyone else). Open the stopcock to release built-up pressure. Repeat this process 2–4 times. Then, close the stopcock and return the funnel upright to the iron ring.

Step 6. Allow the layers to separate and then remove the stopper.

Step 7. Place a suitable clean container just below the tip of the funnel. Gradually open the stopcock and drain the bottom layer into the clean container.

Step 8. Remove the upper layer by pouring it from the top of the funnel. This way it will not become contaminated with traces of the lower layer found in the stem of the funnel.

When aqueous solutions are extracted with a *less dense solvent*, such as ether, the bottom, aqueous layer can be drained *into its original container*. Once the top (organic) layer is removed from the funnel, the aqueous layer can then be returned for further extraction. Losses can be minimized by rinsing the original container with a small portion of the extraction solvent, which is then added to the funnel. When the extraction solvent is denser than the aqueous phase (e.g., methylene chloride), the aqueous phase is the top layer, and therefore is kept in the funnel for subsequent extractions.

CONTINUOUS LIQUID–LIQUID EXTRACTION

Continuous extraction of liquid–liquid systems is also possible and particularly valuable when the component to be separated is only slightly soluble in the extraction solvent. The advantage of using continuous extraction is that it can be carried out with a limited amount of solvent. In batchwise extractions a prohibitive number of individual extractions might have to be performed to accomplish the same overall extraction. Specialized apparatus, however, is required for continuous liquid–liquid extraction.

Two types of continuous extraction apparatus are often used to isolate various species from aqueous solutions using less dense and more dense immiscible solvents (e.g., diethyl ether and methylene chloride) (Fig. 5.18).

The extraction is carried out by allowing the condensate of the extraction solvent, as it forms on the condenser on continuous distillation, to drop through an inner tube (see Fig. 5.18*a* in the case of the less dense solvent) and to percolate up through the solution containing the material to be extracted. This inner tube usually has a sintered glass plug on its end, which generates smaller droplets of the solvent and thus increases the efficiency of the procedure. The extraction solution is then returned to the original distilling flask. Eventually, in this manner, the desired material, extracted in small increments, is collected in the boiling flask and can then be isolated by concentrating the collected solution. This method works on the premise that fresh portions of the less dense phase are continuously introduced into the system, and it is often used in those instances where the organic material to be isolated has an appreciable solubility in water. In the case of a more dense extraction solvent (see Fig. 5.18*b*), the system functions in much the same fashion, but in this case the inner tube is removed and the condensed vapors percolate directly through the lighter phase (the phase to be extracted) to form the lower layer. This layer can cycle back to the distillation flask through a small-bore tubing connection from the bottom of the receiver flask to the distillation flask. Continuous liquid–liquid extraction is useful for removing extractable components from those having partition ratios that approach zero. Note that this method requires a very long period of time.

SEPARATION OF ACIDS AND BASES

The separation of organic acids and bases is another important and extensive use of the extraction method. The distribution coefficients of organic acids and bases are affected by pH when one of the solvents is water. An organic acid that is insoluble in neutral water (pH 7) becomes soluble when the water is made basic with an aqueous sodium hydroxide solution. The acid and the sodium hydroxide quickly react to form a sodium carboxylate salt, RCO_2^-, Na^+. The salt is, of course, ionic, and therefore it readily dissolves in the water.

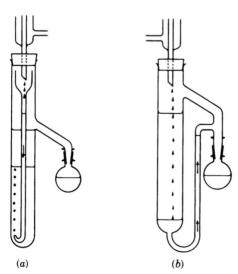

(*a*)   (*b*)

Figure 5.18 Early designs for single-stage extractors: (a) Kutscher–Steudel extractor; (b) Wehrli extractor.

Thus, the acid–base reaction reverses the solubility characteristics of a water-insoluble organic acid.

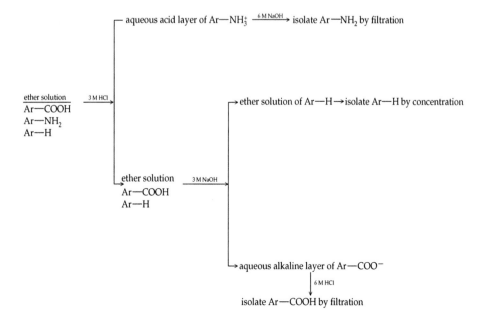

The water phase may then be extracted with an immiscible organic solvent to remove any impurities, leaving the acid salt in the water phase. Neutralizing the water layer with hydrochloric acid (to pH < 7) reprotonates the carboxylate salt to re-form the carboxylic acid, and causes the purified water-insoluble organic acid to precipitate (if it's a solid). In a similar fashion, water-insoluble organic bases, such as amines (RNH_2), can be rendered water soluble by treatment with dilute hydrochloric acid to form water-soluble hydrochloride salts.

Extraction procedures can be used to separate mixtures of solids. For example, the following flow chart diagrams a sequence used to separate a mixture made up of an aromatic carboxylic acid ($ArCO_2H$), an aromatic base ($ArNH_2$), and a neutral aromatic compound (ArH). Aromatic compounds are discussed here simply because they are likely to be crystalline solids.

In this example, we assume that the organic acid and base are solids. If either or both were liquids, an additional extraction of the final acidic aqueous or alkaline solution with ether, followed by drying and concentration, would be required to isolate the acidic or basic component.

SALTING OUT

Most extractions in the organic laboratory involve water and an organic solvent. Many organic compounds have partial solubility in both solvents. To extract them from water, the partition coefficient (between the organic solvent and water) can be shifted in favor of the organic layer by saturating the water layer with an inorganic salt, such as sodium chloride. Water molecules prefer to solvate the polar ions (in this case sodium and chloride ions), and thus free the neutral organic molecules to migrate into the organic phase. Another way to think of this is to realize that the ionic solution is more polar than pure water, so the less polar organic molecules are less soluble than in pure water. Forcing an organic material out of a water solution by adding an inorganic salt is called *salting out*.

Salting out can also be effectively used for the *preliminary drying* of the wet organic layer that results from an extraction process. (Diethyl ether, in particular, can dissolve a fair amount of water.) Washing this organic layer with a saturated salt solution removes most of the dissolved water into the aqueous phase. This makes further drying of the organic phase with solid drying agents easier and much more effective (see the following section, *Drying Agents*).

Solid–liquid extraction

The simplest form of solid–liquid extraction involves treating a solid with a solvent and then decanting or filtering the solvent extract away from the solid. This type of extraction is most useful when only one main component of the solid phase has appreciable solubility in the solvent.

Microscale extractions of trimyristin from nutmeg, and cholesterol from gallstones, have been described.[6] Diethyl ether was used as the solvent in both cases. A packed Pasteur pipet column was used for filtering, drying (nutmeg experiment), and decolorizing (gallstone experiment).

Herrera and Almy described a simple continuous extraction apparatus (Fig. 5.19).[7] The apparatus is constructed from a 50-mL beaker and a paper cone prepared from a 9-cm disk of filter paper (nonfluted), which rests on the lip of the beaker. A small notch is cut in the cone to allow solvent vapor to pass around it. The extraction solvent is placed in the beaker; the solid material to be extracted is placed in the cone. A watch glass containing 2–3 g of ice is placed on top of the assembly to act as the condenser and to hold the paper cone in place. As the ice melts, the water is removed and replaced with fresh ice. The beaker is heated on a hot plate in the hood (some solvent evaporates during the extraction process and may need to be replaced). The concentrated solution collected in the beaker is then cooled, and the solid product is isolated by filtration or is recrystallized. This system needs to be attended at all times, but works reasonably well for brief extractions.

---

[6]Vestling, M. M. *J. Chem. Educ.* **1990**, *67*, 274.
[7]Herrera, A.; Almy, J. *J. Chem. Educ.* **1998**, *75*, 83.

Figure 5.19 A solid–liquid continuous extraction apparatus.

Various apparatus have been developed for use when longer extraction periods are required. They all use what is called a *countercurrent process*. The best-known apparatus is the Soxhlet extractor, first described in 1879 (Fig. 5.20).[8] The solid sample is placed in a porous thimble. The extraction-solvent vapor, generated by refluxing the extraction solvent contained in the distilling pot, passes up through the vertical side tube into the condenser. The liquid condensate then drips onto the solid, which is extracted. The extraction solution

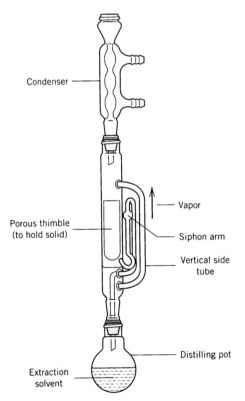

Condenser

Porous thimble
(to hold solid)

Vapor

Siphon arm

Vertical side
tube

Distilling pot

Extraction
solvent

Figure 5.20 Soxhlet extractor.

[8]Soxhlet, F. *Dinglers Polytech. J.* **1879,** 232, 461.

passes through the pores of the thimble, eventually filling the center section of the Soxhlet. The siphon tube also fills with this extraction solution and when the liquid level reaches the top of the tube, siphoning action returns the thimbleful of extract to the distillation pot. The cycle is automatically repeated many times, concentrating the extract in the distillation pot. The advantage of this arrangement is that the process may be continued automatically and unattended for as long as necessary. The solvent is then removed from the extraction solution collected in the pot, providing the extracted compound(s). Soxhlet extractors are available from many supply houses and can be purchased in various sizes. Of particular interest to us is the microscale variety, which is effective for small amounts of material and is now commercially available.[9]

Drying agents

Organic extracts separated from aqueous phases usually contain traces of water. Even washing with saturated salt solution (see Salting Out above) cannot remove all of the water. Organic extracts must therefore be dried to remove any residual water before the solvent is evaporated or further purification is performed. Organic extracts can be conveniently dried with an anhydrous inorganic salt, such as magnesium sulfate, sodium sulfate, or calcium sulfate. These salts readily absorb water and form insoluble hydrates, thus removing the water from the wet organic phase. The hydrated solid can then be removed from the dried solution by filtration or by decanting (pouring) the solution away from the solid. Although many drying agents are known, not every drying agent can be used in every case. The ideal drying agent should dry the solution quickly, have a high capacity for water, cost little, and not react with the material being dried.[10]

Table 5.5 summarizes the properties of some of the more common drying agents used in the laboratory.

Make sure that the solid drying agent is in its *anhydrous* form. Sodium sulfate is a good general-purpose drying agent and is usually the drying agent of choice at room temperature. Use the granular form, if at all possible.

Magnesium sulfate is supplied as a fine powder (high surface area). It has a high water capacity and is inexpensive; it dries solutions more quickly than does sodium sulfate. The disadvantage of magnesium sulfate is that the desired product (or water molecules) can become trapped on the surface of the fine particles. If it is not thoroughly washed after separation, precious product may be lost. Furthermore, it is usually more difficult to remove a finely powdered solid agent, which may pass through the filter paper (if used) or clog the

---

[9]Microscale Soxhlet equipment is available from Ace Glass, Inc., 1430 Northwest Boulevard, Vineland, NJ 08360.

[10]Quantitative studies on the efficiency of drying agents for a wide variety of solvents have been reported. See Burfield, D. R.; Smithers, R. H. *J. Org. Chem.* **1983**, *48*, 2420 and references therein. Other useful information can be found in Armarego, W. L. F.; Perrin, D. D. *Purification of Laboratory Chemicals*, 4th ed.; Butterworth-Heinemann: Woburn, MA, 1996 and in Ridduck, J. A.; Bunger, W. B.; Sakano, T. K. *Organic Solvents, Physical Properties and Methods of Purification*, 4th ed.; Wiley: New York, 1986.

TABLE 5.5 *Properties of Common Drying Agents*

DRYING AGENT	FORMULA OF HYDRATE	COMMENTS
Sodium sulfate	$Na_2SO_4 \cdot 10H_2O$	Slow to absorb water and inefficient, but inexpensive and has a high capacity. Loses water above 32° C. Granular form available.
Magnesium sulfate	$MgSO_4 \cdot 7H_2O$	One of the best. Can be used with nearly all organic solvents. Usually in powder form.
Calcium chloride	$CaCl_2 \cdot 6H_2o$	Relatively fast drying, but reacts with many oxygen- and nitrogen-containing compounds. Usually in granular form.
Calcium sulfate	$CaSO_4 \cdot \frac{1}{2}H_2O$	Very fast and efficient, but has a low dehydration capacity.
Silica gel	$(SiO_2)_m \cdot nH_2O$	High capacity and efficient. Commercially available t.h.e. SiO_2 drying agent is excellent.[a]
Molecular sieves	$[Na_{12}(Al_{12}Si_{12}O_{48})] \cdot 27H_2O$	High capacity and efficient. Use the 4-Å size.[b]

[a]Available from EM Science, Cherry Hill, NJ 08034.
[b]Available from Aldrich Chemical Co. Inc., 940 West Saint Paul Ave., Milwaukee, WI 53233.

pores of a fine porous filter. A smaller surface area translates into less adsorption of product on the surface and easier separation from the dried solution.

Molecular sieves have pores or channels in their structures. A small molecule such as water can diffuse into these channels and become trapped. The sieves are excellent drying agents, have a high capacity, and dry liquids completely. The disadvantages are that they dry slowly and are more expensive than the more common drying agents.

Calcium chloride is very inexpensive and has a high capacity. Use the granular form. Do not use it to dry solutions of alcohols, amines, or carboxylic acids because it can react with these substances.

Calcium sulfate is often sold under the trade name of Drierite. It is a somewhat expensive drying agent. Do not use the blue Drierite (commonly used to dry gases) because the cobalt indicator (blue when dry, pink when wet) may leach into the solvent.

The amount of drying agent needed depends on the amount of water present, on the capacity of the solid desiccant to absorb water, and on its particle size (actually, its surface area). If the solution is wet, the first amount of drying agent will clump (molecular sieves and t.h.e. SiO_2 are exceptions). Add more drying agent until it appears mobile when you swirl the liquid. A solution that is no longer cloudy is a further indication that the solution is dry. Swirling the contents of the container increases the rate of drying; it helps establish the equilibrium for hydration:

$$\text{Drying agent} + n\text{H}_2\text{O} \rightleftharpoons \text{Drying agent} \cdot n\text{H}_2\text{O}$$

Anhydrous solid   Solid hydrate

Most drying agents achieve approximately 80% of their drying capacity within 15 min; longer drying times are generally unnecessary. The drying agent may be added directly to the container of the organic extract, or the extract may be passed through a Pasteur filter pipet packed with the drying agent. A funnel fitted with a cotton, glass wool, or polyester plug to hold the drying agent may also be used.

Solid-phase extraction

In the modern research laboratory, the traditional liquid–liquid extraction technique may be replaced by the solid-phase extraction method.[11] The advantages of this newer approach are that it is rapid, it uses only small volumes of solvent, it does not form emulsions, isolated solvent extracts do not require a further drying stage, and it is ideal for working at the microscale level. This technique is finding wide acceptance in the food industry and in the environmental and clinical area, and it is becoming the accepted procedure for the rapid isolation of drugs of abuse and their metabolites from urine. Solid-phase extraction is accomplished using prepackaged, disposable, extraction columns. A typical column is shown in Figure 5.21. The columns are available from several commercial sources.[12]

The polypropylene columns can be obtained packed with 100–1000 mg of 40-μm sorbent sandwiched between two 20-μm polyethylene frits. The columns are typically 5–6 cm long. Sample volumes are generally 1–6 mL.

The adsorbent (stationary phase) used in these columns is a nonpolar adsorbent chemically bonded to silica gel. In fact, they are the same nonpolar adsorbents used in the reversed-phase high-performance liquid chromatography (HPLC). More specifically, the adsorbents are derivatized silica gel where the —OH groups of the silica gel have been replaced with siloxane groups by treating silica gel with the appropriate organochlorosilanes.

Silica surface   Chemically bonded silica surface

[11]For a description of this method, see Zief, M.; Kiser, P. *Am. Lab.* **1990,** 70; Zief, M. *NEACT J.* **1990,** *8,* 38; Hagen, D. F.; Markell, C. G.; Schmitt, G. A.; Blevins, D. D. *Anal. Chim. Acta* **1990,** *236,* 157; Arthur, C. L.; Pawliszyn, J. *Anal. Chem.* **1990,** *62,* 2145; Junk, G. A.; Richard, J. J. *Anal. Chem* **1988,** *60,* 451 and references therein; Dorsey, J.; Dill, K. A. *Chem. Rev.* **1989,** *89,* 331; Nawrocki, J.; Baszewski, B. *J. Chromatogr.* **1988,** *559,* 1; Cooke, N.H.C.; Olsen, K. *Am. Lab.* **1979,** *11,* 45; Baker-10 Special Applications Guide, Vol. 1, p. 182, J. T. Baker, Inc., Phillipsburg, NJ.

[12]These columns are available from Analytichem International, J. T. Baker, Inc., Supelco, Inc., Aldrich Chemical, Waters Associates, and Biotage (a Division of Dyax Corp.).

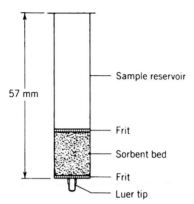

Figure 5.21 Polyethylene solid-phase extraction column.

— Sample reservoir

— Frit

— Sorbent bed

— Frit

— Luer tip

57 mm

Two of the most popular nonpolar packings are those containing R groups consisting of an octadecyl ($C_{18}H_{37}$—) or phenyl (C_6H_5—) group. These packing materials (stationary phases) can adsorb nonpolar (like attracts like) organic material from aqueous solutions. The adsorbed material is then eluted from the column using a solvent strong (nonpolar) enough to displace it, such as methanol, methylene chloride, or hexane. The analyte capacity of bonded silica gels is about 10–20 mg of analyte per gram of packing.

An example of a typical solid-phase extraction is the determination of the amount of caffeine in coffee using a 1-mL column containing 100 mg of octadecyl-bonded silica. This efficient method isolates about 95% of the available caffeine. The column is conditioned by flushing 2 mL of methanol followed by 2 mL of water through the column. One milliliter of a coffee solution (~0.75 mg of caffeine/mL) is then drawn through the column at a flow rate of 1 mL/min. The column is washed with 1 mL of water and air dried (vacuum) for 10 min. The adsorbed caffeine is then eluted with two 500-μL portions of chloroform.

QUESTIONS

5-30. If, when extracting an aqueous solution with an organic solvent, you were uncertain which layer was the organic one, how could you quickly determine which was which?

5-31. Which layer (upper or lower) will each of the following organic solvents usually form when being used to extract an aqueous solution?
    toluene   methylene chloride   diethyl ether
    hexane   acetone

5-32. Construct a flow chart to demonstrate how you could separate a mixture of 1,4-dichlorobenzene, 4-chlorobenzoic acid, and 4-chloroaniline using an extraction procedure.

5-33. A slightly polar organic compound partitions itself between ether and water phases. The K_p (partition coefficient) value is 2.5 in favor of the ether solvent. What simple procedure could you use to increase this K_p value?

5-34. You weight out exactly 1.00 mg of benzoic acid and dissolve it in a mixture of 2.0 mL of diethyl ether and 2.0 mL of water. After mixing and allowing the layers to separate, the ether layer is removed, dried, and concentrated to yield 0.68 mg of benzoic acid. What is the K_p value (ether/water) for this system?

BIBLIOGRAPHY

For overviews on extraction methods see the following general references:

CRAIG, L. C.; CRAIG, D. In *Techniques of Organic Chemistry*, 2nd ed.; Weissberger, A. Ed.; Interscience, New York, 1956; Vol. III, Part I, Chapter 2.

JUBERMANN, O. In *Houben-Weyl Methoden der Organischen Chemie*, 4th ed.; Verlag: Stuttgart, 1958, Vol. I, p. 223.

Kirk–Othmer Encyclopedia of Chemical Technology, 4th ed.; Wiley: New York, 1993; Vol. 10, p. 125.

SCHNEIDER, FRANK L. *Qualitative Organic Microanalysis*, Vol. II of *Monographien aus dem Gebiete der qualitativen Mikroanalysis*, A. A. Benedetti-Pichler, Ed.; Springer-Verlag: Vienna, 1964, p. 61.

SHUGAR, G. J. *Chemical Technicians' Ready Reference Handbook*, 3rd ed.; McGraw-Hill: New York, 1990.

THORPE, J. F.; WHITELEY, M. A. *Thorpe's Dictionary of Applied Chemistry*, 4th ed.; Longmans: New York, 1940; Vol. IV, p. 575.

ZUBRICK, J. W. *The Organic Chem Lab Survival Manual: A Student's Guide to Techniques*, 4th ed.; Wiley: New York, 1997.

TECHNIQUE 5
Crystallization

This discussion introduces the basic technique of purifying solid organic substances by crystallization. The technique of crystallizing an organic compound is fundamental; it must be mastered if you are going to purify solids. *It is not an easy art to acquire.* Organic solids tend not to crystallize as easily as inorganic substances.

Legend has it that an organic chemist resisted an invitation to leave a well-worn laboratory for new quarters because he suspected that the older laboratory (in which many crystallizations had been carried out) harbored seed crystals for a large variety of substances the chemist needed. Carried by dust from the earlier work, these traces of material presumably aided the successful initiation of crystallization of reluctant materials. Further support for this legend comes from the often quoted (but never substantiated) belief that after a material was first crystallized in a particular laboratory, subsequent crystallizations of the material, regardless of its purity or origin, were always easier.

In several areas of chemistry, the success or failure of an investigation can depend on the ability of a chemist to isolate tiny quantities of crystalline substances. Often the compounds of interest must be extracted from enormous amounts of extraneous material. In one of the more spectacular examples, Reed et al. isolated 30 mg of the crystalline coenzyme lipoic acid from 10 tons of beef liver residue.[13]

$$\text{S—S} \quad CH_2CH_2CH_2CH_2CO_2H$$
$$H$$

Lipoic acid

---

[13]Reed, L. J.; Gunsalus, I. C.; Schnakenberg, G. H. F.; Soper, Q. F.; Boaz, H. E.; Kem, S. F.; Parke, T. V. *J. Am. Chem. Soc.* **1953,** 75, 1267.

General crystallization procedure

The following steps are the essentials of crystallization:

Step 1. Select a suitable solvent.

Step 2. Dissolve the material to be purified in the minimum amount of warm solvent. *Remember that most organic solvents are extremely flammable and that many produce very toxic vapor.*

Step 3. Once the solid mixture has fully dissolved, filter the heated solution, and then bring it to the point of saturation by evaporating a portion of the solvent.

Step 4. Cool the warm saturated solution to reduce the solubility of the solute; this usually causes the solid material to precipitate. If the material has a low melting point or is very impure it may come out of solution sometimes as an oil. If so, reheat the solution and allow it to recool slowly.

Step 5. Isolate the solid by filtration, and then remove the last traces of solvent.

The crystallization is successful if the solid is recovered in good yield and is purer than it was before the crystallization. This cycle, from solid state to solution and back to solid state is called *recrystallization* when both the initial and final materials are crystalline.

Although the technique sounds fairly simple, in reality it is demanding. Successful purification of microscale quantities of solids will require your utmost attention. Choosing a solvent system is critical to a successful crystallization. To achieve high recoveries, the compound to be crystallized should ideally be very soluble in the hot solvent, but nearly insoluble in the cold solvent. To increase the purity of the compound, the impurities should be either *very soluble* in the solvent at all temperatures or *not soluble* at any temperature. The solvent should have as low a boiling point as possible so that traces of solvent can be easily removed (evaporated) from the crystals after filtration. It is best to use a solvent that has a boiling point at least 10 °C lower than the melting point of the compound to be crystallized to prevent the solute from "oiling out" of the solution. Thus, the choice of solvent is critical to a good crystallization. Table 5.6 lists common solvents used in the purification of organic solids.

When information about a suitable solvent is not available, the choice of solvent is made on the basis of solubility tests. Craig's rapid and efficient procedure for microscale solubility testing works nicely; it requires only milligrams of material and a nine-well, Pyrex spot plate.[14]

• Place 1–2 mg of the solid in each well and pulverize each sample with a stirring rod. Add 3–4 drops of a given solvent to the first well and observe whether the material dissolves at ambient temperature. If not, stir the mixture for 1.5–2 min and observe and record the results. Repeat this process with the chosen set of solvents, using a separate well for each solubility test. *Keep track of which well contains which solvent.* Place your test plate (containing the samples) on a hot plate (set at its lowest setting) in the **hood**; add additional solvent if necessary. Record the solubility characteristics of the sample in each hot

[14]Craig, R. E. R. *J. Chem. Educ.* **1989,** *66,* 88.

TABLE 5.6 *Common Solvents*

SOLVENT	bp (°C)	POLARITY
Acetone	56	Polar
Cyclohexane	81	Nonpolar
Diethyl ether	35	Intermediate polarity
Ethanol, 95%	78	Polar
Ethyl acetate	77	Intermediate polarity
Hexane	68	Nonpolar
Ligroin	60–90	Nonpolar
Methanol	65	Polar
Methylene chloride	40	Intermediate polarity
Methyl ethyl ketone	80	Intermediate polarity
Petroleum ether	30–60	Nonpolar
Toluene	111	Nonpolar
Water	100	Polar

solvent. Cool the plate and see if crystallization occurs in any of the wells. On the basis of your observations, choose an appropriate solvent or a solvent pair (see the following paragraph) to recrystallize your material.

Solubility relationships are seldom ideal for crystallization; most often a compromise is made. If there is no suitable single solvent available, it is possible to use a mixture of two solvents, called a *solvent pair*. A solvent is chosen that will readily dissolve the solid. Once the solid is dissolved in the minimum amount of hot solvent, the solution is filtered. A second solvent, miscible with the first, in which the solute has much lower solubility, is then added dropwise to the hot solution to achieve saturation. In general, polar organic molecules have higher solubilities in polar solvents, and nonpolar materials are more soluble in nonpolar solvents (like dissolves like). Table 5.7 lists some common solvent pairs.

It can take a long time to work out an appropriate solvent system for a particular reaction product. In most instances, with known compounds, the optimum solvent system has been established. Most crystallizations are not very efficient because many impurities have solubilities similar to those of the compounds of interest. Recoveries of 50–70% are not uncommon.

Several microscale crystallization techniques are available.

Simple crystallization

Simple crystallization works well with large quantities of material (100 mg and up), and it is essentially identical to that of the macroscale technique.

Step 1. Place the solid in a small Erlenmeyer flask or test tube. A beaker is not recommended because the rapid and dangerous loss of flammable vapors of hot solvent occurs much more easily from the wide mouth of a beaker than from an Erlenmeyer flask. Furthermore, solid precipitate can rapidly collect on the walls of the beaker as the solution becomes saturated because the atmosphere above the solution is less likely to be saturated with solvent vapor in a beaker than in an Erlenmeyer flask.

TABLE 5.7 *Common Solvent Pairs*

SOLVENT 1(MORE POLAR)	SOLVENT 2(LESS POLAR)
Acetone	Diethyl ether
Diethyl ether	Hexane
Ethanol	Acetone
Ethyl acetate	Cyclohexane
Methanol	Methylene chloride
Acetone	Water
Water	Ethanol
Toluene	Ligroin

Step 2. Add a minimal amount of solvent and heat the mixture to the solvent's boiling point in a sand bath. Stir the mixture by twirling a spatula between the thumb and index finger. A magnetic stir bar may be used if a magnetic stirring hot plate is used.

Step 3. Continue stirring and heating while adding solvent dropwise until all of the material has dissolved.

Step 4. Add a decolorizing agent (powdered charcoal, ~2% by weight; or better, activated-carbon Norit pellets,[15] ~0.1% by weight), to remove colored minor impurities and other resinous byproducts.

Step 5. Filter (by gravity) the hot solution into a second Erlenmeyer flask (preheat the funnel with hot solvent). This removes the decolorizing agent and any insoluble material initially present in the sample.

Step 6. Evaporate enough solvent to reach saturation.

Step 7. Cool to allow crystallization (crystal formation will be better if this step takes place slowly). After the system reaches room temperature, cooling it in an ice bath may improve the yield.

Step 8. Collect the crystals by filtration on a Büchner or Hirsch funnel. Save the mother liquor (this is the term used to describe the solution that was separated from the original crystals) until the identity of the product has been established. In some cases, it is possible to recover more product by concentrating and further cooling the mother liquor. The second crop of crystals, however, is usually not as pure as the first.

Step 9. Wash (rinse) the crystals carefully.

Step 10. Dry the crystals.

Filtration techniques

USE OF THE HIRSCH FUNNEL

The standard filtration system for collecting products purified by recrystallization in the microscale laboratory is vacuum (suction) filtration with an

---

[15]Available from Aldrich Chemical Co., 940 West St. Paul Ave., Milwaukee, WI 53233, or http://www.sigma-aldrich.com. Catalog No. 32942-8.

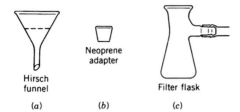

Hirsch funnel

(a)

Neoprene adapter

(b)

Filter flask

(c)

Figure 5.22 Component parts for vacuum filtration.

11-mm Hirsch funnel. Many reaction products that do not require recrystallization can also be collected directly by vacuum filtration. The Hirsch funnel (Fig. 5.22a) is composed of a ceramic cone with a circular flat bed perforated with small holes. The diameter of the bed is covered by a flat piece of filter paper of the same diameter. The funnel is sealed into a filter flask with a Neoprene adapter (Fig. 5.22b). Plastic and glass varieties of this funnel that have a polyethylene or glass frit are now available. It is still advisable to use the filter paper disk with these funnels to prevent the frit from clogging or becoming discolored. Regardless of the type of filter used, *always wet the filter paper disk with the solvent being used in the crystallization and then apply the vacuum*. This ensures that the filter paper disk is firmly seated on the bed of the filter.

Filter flasks have thick walls, and a side arm to attach a vacuum hose, and are designed to operate under vacuum (see Fig. 5.22c). The side arm is connected with *heavy-walled* rubber vacuum tubing to a water aspirator (water pump). The water pump uses a very simple aspirator based on the Venturi effect. Water is forced through a constricted throat in the pump. [See the detailed discussion of the Venturi effect and water pumps in the section on reduced pressure (vacuum) distillations]. When water flows through the aspirator, the resulting partial vacuum sucks air down the vacuum tubing from the filter flask. *Always turn the water on full force*. With the rubber adapter in place, air is pulled through the filter paper, which is held flat on the bed of the Hirsch (or Büchner) funnel by the suction. The mother liquors are rapidly forced into the filter flask, where the pressure is lower, by atmospheric pressure. The crystals retained by the filter are dried by the stream of air passing through them (Fig. 5.23).

When you use a water pump, *it is very important to have a safety trap mounted in the vacuum line leading from the filter flask*. Any drop in water pressure at the pump (easily created by other students on the same water line turning on other aspirators at the same time) can result in a backup of water into the flask. As the flow through the aspirator decreases, the pressure at that point rapidly increases and water is forced up the vacuum tubing toward the filter flask (Fig. 5.23). It is also important to vent the vacuum by opening the vent stopcock or

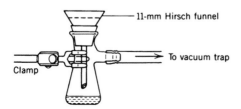

11-mm Hirsch funnel

To vacuum trap

Clamp

Figure 5.23 Vacuum filtration apparatus.

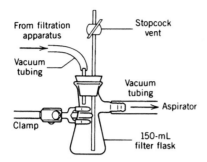

Figure 5.24 Vacuum trap.

disconnecting the rubber tubing from the filter flask, *before* the water is turned off (see Fig. 5.24).

In some cases, the precipitate collected on the Hirsch funnel is not highly crystalline. The filter cake may be too thick or pasty to dry simply by pulling air through it. A thin, flexible rubber sheet or a piece of plastic food wrap placed over the mouth of the funnel, such that the suction generated from the vacuum pulls the sheeting down onto the filter cake (collected crystals), will place atmospheric pressure on the solid cake. This pressure can force much of the remaining solvent from the collected material, and thus further dry it. Use a piece of sheeting large enough to cover the entire filter cake. Otherwise, a vacuum may not be created and adequate drying may not occur.

In some instances, substances may retain water or other solvents with great tenacity. To dry these materials, a *desiccator* is often used. This is generally a glass or plastic container containing a *desiccant* (a material capable of absorbing water). The substance to be dried, held in a suitable container, is then placed on a support above the desiccant. This technique is often used in quantitative analysis to dry collected precipitates. Vacuum desiccators are available (Fig. 5.25a). If this method of drying is still insufficient, a *drying pistol* can be used (Fig. 5.25b). The sample, in an open container (vial) is placed in the apparatus, which is then evacuated. The pistol has a pocket in which a strong adsorbing agent, such as P_4O_{10} (for water), NaOH or KOH (for acidic gases), or

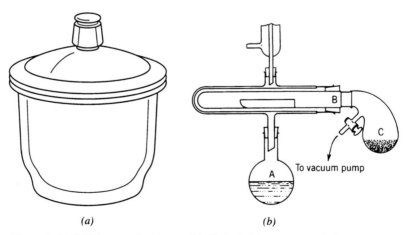

(a) (b)

Figure 5.25 (a) Vacuum desiccator. (b) Abderhalden vacuum drying apparatus. *A*, refluxing heating liquid; *B*, vacuum drying chamber; *C*, desiccant.

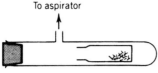

(20 x 150 mm sidearmed test tube)

Figure 5.26 Side-arm test tube as a vacuum drying apparatus.

paraffin wax (for organic solvents), is placed. The pistol is heated by refluxing vapors that surround the barrel. A simple alternative to this method is the use of a side-armed test tube (Fig. 5.26).

A HIRSCH-FUNNEL ALTERNATIVE—A NAIL-FILTER FUNNEL

A nail-filter funnel is a low-cost substitute for a Hirsch funnel.[16] This apparatus is easily assembled from common laboratory glassware. Obtain a soft-glass rod that fits in the stem of a small glass funnel. Cut the rod to a suitable size and heat the tip of one end over a burner flame. When the tip becomes soft, hold the rod vertically and press the hot tip against a cold metal surface to flatten it to form a flat nail-like head. *The nail head should not be perfectly round or it will block the flow of liquid.* Cut the cooled rod to a suitable length and place the "nail" inside the stem of the funnel so that the flattened head of the nail rests on the top opening of the funnel. Cut a piece of filter paper just slightly larger than the nail head, place it on the nail head, and then place the funnel in a filter flask with a neoprene adapter. Be sure to wet the filter paper before filtering.

CRAIG TUBE CRYSTALLIZATIONS

The Craig tube[17] is commonly used for microscale crystallizations in the range of 10–100 mg of material (Fig. 5.27). The process consists of the following steps.

Step 1. Place the sample in a small test tube (10 × 75 mm).

Step 2. Add the solvent (0.5–2 mL), and dissolve the sample by heating in the sand bath; add drops of solvent as needed. Rapid stirring with a microspatula (roll the spatula rod between your thumb and index finger) helps dissolve the material and protects against boilover. Add several drops of solvent by Pasteur pipet after the sample has completely dis-

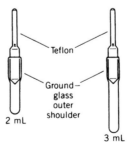

Teflon

Ground–glass outer shoulder

2 mL

3 mL **Figure 5.27** Craig tubes.

---

[16]Szafran, Z.; Pike, R. M.; Singh, M. M. *Microscale & Selected Macroscale Experiments for General & Advanced General Chemistry;* Wiley: New York, 1995, pp. 47, 63; Claret, P. A. *Small Scale Organic Preparations;* Pitman: London, 1961, p. 15.

[17]Craig, L. C.; Post, O. W. *Ind. Eng. Chem., Anal. Ed.* **1944,** *16,* 413.

solved. It will be easy to remove this excess at a later stage, since the volumes involved are very small. The additional solvent ensures that the solute will stay in solution during the hot transfer. Norit charcoal pellets may be added at this stage, if needed to remove colored impurities.

Step 3. Transfer the heated solution to the Craig tube by Pasteur filter pipet (see Fig. 3.20, p. 33) that has been preheated with hot solvent. This transfer automatically filters the solution. A second filtration is often necessary if powdered charcoal has been used to decolorize the solution.

Step 4. The hot, filtered solution is then concentrated to saturation by gentle boiling in the sand bath. Constant agitation of the solution with a microspatula during this short period can avoid the use of a boiling stone and prevent boilover. Ready crystallization on the microspatula just above the solvent surface is a good indication that saturation is close at hand.

Step 5. The upper section of the Craig tube (the "head" or stopper) is set in place, and the saturated solution is allowed to cool in a safe place. As cooling commences, seed crystals, if necessary, may be added by crushing them against the side of the Craig tube with a microspatula just above the solvent line. A good routine, if time is available, is to place the assembly in a small Erlenmeyer, then place the Erlenmeyer in a beaker, and finally cover the first beaker with a second inverted beaker. This procedure will ensure slow cooling, which will enhance good crystal growth (Fig. 5.28). A Dewar flask may be used when very slow cooling and large crystal growth are required.

Step 6. After the system reaches room temperature, cooling in an ice bath may improve the yield.

Step 7. Remove the solvent by inverting the Craig tube assembly into a centrifuge tube and spinning the mother liquors away from the crystals (Fig. 5.29). *This operation should be carried out with care.* First, fit the head with a thin copper wire (Fig. 5.29), held in place by a loop at the end of the wire that is placed around the narrow part of the neck. Some Teflon heads have a hole in the neck to anchor the wire. The copper wire should not be much longer than the centrifuge tube.

Step 8. Now insert the Craig tube into a centrifuge tube. To do this, hold the Craig tube upright (with the head portion up) and place the centrifuge tube *down over* the Craig tube. Push the Craig tube up with your finger

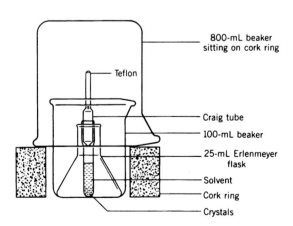

800-mL beaker
sitting on cork ring

Teflon

Craig tube

100-mL beaker

25-mL Erlenmeyer
flask

Solvent

Cork ring

Crystals

Figure 5.28 Apparatus for slow crystallization.

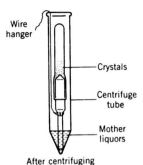

Figure 5.29 Crystal collection with a Craig tube.

so that the head is against the inverted bottom of the centrifuge tube, and then invert the whole assembly (Fig. 5.29).

Step 9. Place the assembly into a centrifuge tube, *balance the centrifuge,* and spin the mother liquors away from the crystals (Fig. 5.29). This replaces the usual filtration step in simple crystallizations. It avoids another transfer of material and also avoids product contact with filter paper.

Step 10. Remove the apparatus from the centrifuge, and then carefully remove the Craig tube from the centrifuge tube. Gently pull upward on the copper wire while at the same time applying downward pressure with your fingers to the bottom of the inverted Craig tube (this will keep the Craig tube assembly together and not let any of the crystalline product fall back into the centrifuge tube and into the mother liquors). Once the Craig tube assembly is removed from the centrifuge tube, turn it so that the neck of the tube is up, and then disassemble it. At this point scrape any crystalline product clinging to the head into the lower section. If the lower section is tared, it can be left to air dry to constant weight or placed in a warm vacuum oven (use a rubber band or thin wire to wrap a piece of filter paper over the open end to prevent dust from collecting on the product while drying).

The cardinal rule in carrying out the purification of small quantities of solids is *Keep the number of transfers to an absolute minimum!* The Craig tube is very helpful in this regard.

For other recrystallization and filtration methods, see 🌐; Chapter 5, Crystallization.

QUESTIONS

5-35. What is the purpose of using activated carbon in a recrystallization procedure?

5-36. List several advantages and disadvantages of using a Craig tube for recrystallization.

5-37. Why is it advisable to use a stemless or a short-stemmed funnel when carrying out a gravity filtration?

5-38. Which of the following solvent pairs could be used in a recrystallization? Why or why not?

(a) Acetone and ethanol

(b) Hexane and water

(c) Hexane and diethyl ether

5-39. You perform a recrystallization on 60 mg of a solid material and isolate 45 mg of purified material. What is the percent recovery? Further concentration of the mother liquor provides an additional 8 mg of material. What is the total percent recovery?

5-40. Describe two techniques that can be used to induce crystallization.

5-41. When would you advise someone to use a solvent pair to carry out a recrystallization?

TECHNIQUE 6
Chromatography

TECHNIQUE 6A
Column, Flash, High-Performance Liquid, and Thin-Layer Chromatography

The basic theory of chromatography is introduced in Technique 1 in the discussion of gas-phase separations. The word *chromatography* is derived from the Greek word for color, *chromatos.* Tswett discovered the technique in 1903 while studying ways to separate mixtures of natural plant pigments.[18] The chromatographic zones were detected simply by observing the visual absorption bands. Thus, as originally applied, the name was not an inconsistent use of terminology. Today, however, most mixtures are colorless. The separated zones in these cases are detected by other methods.

Two chromatographic techniques are discussed in this section. Both depend on adsorption and distribution between a stationary solid phase and a moving liquid phase. The first is column chromatography, which is used extensively throughout organic chemistry. It is one of the oldest of the modern chromatographic methods. The second technique, thin-layer chromatography (TLC), is particularly effective in rapid assays of sample purity. It can also be used as a preparative technique for obtaining tiny amounts of high-purity material for analysis.

Column chromatography

Column chromatography, as its name implies, uses a column packed with a solid stationary phase. A mobile liquid phase flows by gravity (or applied pressure) through the column. Column chromatography uses polarity differences to separate materials. A sample on a chromatographic column is subjected to two opposing forces: (1) the solubility of the sample in the elution solvent system, and (2) the adsorption forces binding the sample to the solid phase. These

---

[18]Tswett, M. *Ber. Deut. Botan. Ges.* **1906,** 24, 235.

interactions comprise an equilibrium. Some sample constituents are adsorbed more tightly; other components of the sample dissolve more readily in the liquid phase and are eluted more rapidly. The more rapidly eluting materials, thus, are carried further down the column before becoming readsorbed, and thus exit the column before more tightly bound components. The longer the column, the larger the number of adsorption–dissolution cycles (much like the vaporization–condensation cycles in a distillation column), and the greater the separation of sample components as they elute down the column. A molecule that is strongly adsorbed on the stationary phase will move slowly down the column; a molecule that is weakly adsorbed will move at a faster rate. Thus, a complex mixture can be resolved into separate bands of pure materials. These bands of purified material eventually elute from the column and can be collected.

Many materials have been used as the stationary phase in column chromatography. Finely ground alumina (aluminum oxide, Al_2O_3) and silicic acid (silica gel, SiO_2) are by far the most common adsorbents (stationary phases). Many common organic solvents are used as the liquids (sometimes called *eluents*) that act as the mobile phase and elute (wash) materials through the column. Table 5.8 lists the better known column packing and elution solvents.

Silica gel impregnated with silver nitrate (usually 5–10%) is also a useful adsorbent for some functional groups. The silver cation selectively binds to unsaturated sites via a silver-ion π complex. Traces of alkenes are easily removed from saturated reaction products by chromatography with this system. This adsorbent, however, must be protected from light until used, or it will quickly darken and become ineffective.

Column chromatography is usually carried out according to the procedures discussed in the following five sections.

PACKING THE COLUMN

The quantity of stationary phase required is determined by the sample size. A common rule of thumb is to use a weight of packing material 30–100 times the weight of the sample to be separated. Columns are usually built with roughly a 10:1 ratio of height to diameter. In the microscale laboratory, two standard chromatographic columns are used:

TABLE 5.8 *Column Chromatography Materials*

STATIONARY PHASE		MOVING PHASE	
Alumina	*Increasing adsorption of polar materials* ↑	Water	*Increasing solvation of polar materials* ↑
Silicic acid		Methanol	
Magnesium sulfate		Ethanol	
Cellulose paper		Acetone	
		Ethyl acetate	
		Diethyl ether	
		Methylene chloride	
		Cyclohexane	
		Pentane	

1. A Pasteur pipet, modified by shortening the capillary tip, is used to separate smaller mixtures (10–100 mg). Approximately 0.5–2.0 g of packing is used in the pipet column (Fig. 5.30*a*).

2. A 50-mL titration buret (modified by shortening the column to 10 cm above the stopcock) is used for larger (50–200 mg) or difficult-to-separate sample mixtures. A buret column uses approximately 5–20 g of packing (Fig. 5.30*b*).

Both columns are prepared by first clamping the empty column in a vertical position and then seating a small cotton or glass wool plug at the bottom. For a buret column, cover the cotton with a thin layer of sand. The Pasteur pipets are loaded by adding the dry adsorbent with gentle tapping, "dry packing." The pipet column (*dry column*) is then premoistened just prior to use.

The burets (*wet columns*) are packed by a slurry technique. In this procedure the column is filled part way with solvent; then the stopcock is opened slightly, and as the solvent slowly drains from the column a slurry of the adsorbent–solvent is poured into the top of the column. The column should be gently tapped while the slurry is added. The solvent is then drained to the top of the adsorbent level and held at that level until used. Alternatively, the wet-packed column can be loaded by sedimentation techniques rather than using a slurry. One such routine is to initially fill the column with the least-polar solvent to be used in the intended chromatographic separation. Then the solid phase is slowly added with gentle tapping, which helps to avoid subsequent channeling. As the solid phase is added, the solvent is slowly drained from the buret at the same time. After the adsorbent has been fully loaded, the solvent level is then lowered to the top of the packing as in the slurry technique.

SAMPLE APPLICATION

Using a Pasteur pipet, apply the sample in a minimum amount of solvent (usually the least polar solvent in which the material is readily soluble) to the

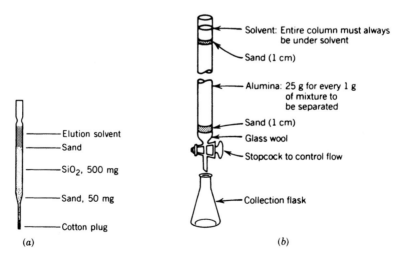

Figure 5.30 Chromatographic columns: (a) a Pasteur pipet column; (b) a buret column. *(From Zubrick, James W. The Organic Chem Lab Survival Manual, 4th ed.; Wiley: New York, 1997. Reprinted by permission of John Wiley & Sons, Inc., New York.)*

top of the column. *Do not disturb the sand layer!* Rinse the pipet, and add the rinses to the column just as the sample solution drains to the top of the adsorbent layer.

ELUTION OF THE COLUMN

The critical step in resolving the sample mixture is eluting the column. Once the sample has been applied to the top of the column, the elution begins (a small layer of sand can be added to the top of the buret column after addition of the first portion of elution solvent).

Note

> Do not let the column run dry: This can cause channels to form in the column.

In a buret column, the flow is controlled by the stopcock. The flow rate should be set to allow time for equilibrium to be established between the two phases; this will depend on the nature and amount of the sample, the solvent, and how difficult the separation will be. The Pasteur pipet column is *free flowing* (the flow rate is controlled by the size of the capillary tip and its plug); once the sample is on the column, the chromatography will require constant attention.

If necessary, it is possible to ease this restriction somewhat by modifying the pipet. Place a Tygon connector (short sleeve) at the top of the pipet column. Once the sample is on the column, insert a second pipet into this connector with its tip just below the liquid level on the top of the column. Add additional solvent through the second pipet (use a bulb, if necessary, but remove it before the elution begins), which acts as a solvent reservoir. As the solvent level in the column pipet drops below the tip of the top pipet, air is admitted, and additional solvent is automatically delivered to the chromatographic column. Thus, the solvent head on the column is maintained at a constant volume. The top pipet need be filled only at necessary intervals; larger volumes of solvent can thus be added to this reservoir. This arrangement also prevents dislodging of the absorbent as new solvent is added.

Note

> It is exceedingly important that solvents do not come in contact with the Tygon sleeve holding the second pipet. These sleeves contain plasticizers that will readily dissolve and contaminate the sample.

The choice of solvent is dictated by a number of factors. A balance between the adsorption power of the stationary phase and the solvation power of the elution solvent governs the rate of travel of the material down the column. If the material travels rapidly down the column, then too few adsorption–elution cycles will occur and the materials will not separate. If the sample travels too slowly, diffusion broadening takes over and separation is degraded. Solvent choices and elution rates can strike a balance between these factors and maximize the separation. It can take considerable time to develop a solvent or mixture of solvents that produces a satisfactory separation of a particular mixture.

FRACTION COLLECTION

As the solvent elutes from the column, it is collected in a series of "fractions" by using small Erlenmeyer flasks or vials. Under ideal conditions, as the mixture of materials travels down the column, it will separate into several individual bands (zones) of pure substances. By careful collection of the fractions, these bands can be separated as they sequentially elute from the column (similar to the collection of GC fractions in the example described in Technique 1). The bands of eluted material can be detected by a number of techniques (weighing fraction residues, colored materials, TLC, etc.).

Column chromatography is a powerful technique for the purification of organic materials. In general, it is significantly more efficient than crystallization procedures. Recrystallization is often best avoided until the last stages of purification, when it will be most efficient. Rely instead on chromatography to do most of the separation.

Column chromatography of a few milligrams of product usually takes no more than 30 min, but chromatographing 10 g of product might easily take several hours, or even all day. Large-scale column chromatography (50–100 g) of a complex mixture could take several days to complete using this type of equipment.

Flash chromatography

Flash chromatography, first described by Still and co-workers in 1978, is a common method for separating and purifying nonvolatile mixtures of organic compounds.[19] The technique is rapid, easy to perform, relatively inexpensive, and gives good separations. Many laboratories routinely use flash chromatography to separate mixtures ranging from 10 mg to 10 g.

This moderate-resolution, preparative chromatography technique was originally developed using silica gel (40–63 μm). Bonded-phase silica gel of a larger particle size can be used for reversed-phase flash chromatography. Flash chromatography columns are generally packed dry to a height of approximately 6 in. Thin-layer chromatography is a quick way to choose solvents for flash chromatography. A solvent that gives differential retardation factor (DR_f) values (of the two substances requiring separation) ≥ 0.15 on TLC usually gives effective separation with flash chromatography. Table 5.9 lists typical experimental parameters for various sample sizes, as a guide to separations using flash chromatography. In general, a mixture of organic compounds separable by TLC can be separated preparatively using flash chromatography.

---

[19]Still, W. C.; Kahn, M.; Mitra, A. *J. Org. Chem.* **1978**, *43*, 2923.

TABLE 5.9 *Typical Experimental Parameters*

COLUMN DIAMETER (mm)	TOTAL VOLUME OF ELUENT (mL)[a]	TYPICAL SAMPLE LOADING (mg)		TYPICAL FRACTION (mL)
		$DR_{f > 0.2}$	$DR_{f > 0.1}$	
10	100–150	100	40	5
20	200–250	400	160	10
30	400–450	900	360	20
40	500–550	1600	600	30
50	1000–1200	2500	1000	50

Source. Data from Majors, R. E., and Enzweiller, T. *LC, GC,* **1998,** *6,* 1046.
[a]Required for both packing and elution.

A flash chromatography apparatus generally consists of a glass column equipped to accept a positive pressure of compressed air or nitrogen applied to the top of the column. A typical commercially available arrangement is shown in Figure 5.31.[20]

Generally, a 20–25% solution of the sample in the elution solvent is recommended, as is a flow rate of about 2 in./min. The column must be conditioned, before the sample is applied, by flushing the column with the elution solvent to drive out air trapped in the stationary phase and to equilibrate the stationary phase and the solvent.

Several modifications of the basic arrangement have been reported, especially in regard to the adaptation of the technique to the instructional laboratory. These involve inexpensive pressure control valves, use of an aquarium "vibrator" air pump, and adapting a balloon reservoir to supply the pressurized gas.

At the microscale level, a pipet bulb or pump on the pipet column can be used to supply pressure to the column. If a bulb is used, squeeze it to apply pressure, and *remove the bulb from the pipet before releasing it!* Otherwise, material may be sucked up into the bulb and most likely disturb the column. Reapply the bulb to re-create pressure. If a pump is used, do not back off the pressure once it has been applied.

An improved method, utilizing a capillary Pasteur pipet for introducing the sample onto the chromatographic column approximately doubles the effectiveness (theoretical plates) of the column.[21] Dry-column flash chromatography[22] has been adapted for use in the instructional laboratory.[23] The "column" consists of a dry bed of silica gel in a sintered glass funnel placed in a standard vacuum filtration flask; the solvent is eluted by suction. Small (16 × 150-mm) test tubes inserted into the flask below the stem of the funnel are used

---

[20]A complete line of glass columns, reservoirs, clamps, and packing materials for flash chromatography is offered by Aldrich Chemical Co., P.O. Box 355, Milwaukee, WI 53201, or http://www.sigma-aldrich.com. Silica gels for use in this technique are also available from Amicon, Danvers, MA; J. T. Baker, Phillipsburg, NJ; EM Science/Merck, Gibbstown, NJ.; ICN Biomedicals, Inc., Cleveland, OH; Universal Solvents, Atlanta, GA; Whatman, Clifton, NJ.

[21]Pivnitsky, K. K. *Aldrichimica Acta* **1989,** *22,* 30.

[22]Harwood, L. M.; *Aldrichimica Acta* **1985,** *18,* 25; Sharp, J. T.; Gosney I.; Rowley A. G. *Practical Organic Chemistry*; Chapman & Hall: New York, 1989.

[23]Shusterman, A. J.; McDougal, P. G.; Glasfeld, A. *J. Chem. Educ.* **1997,** *74,* 1222.

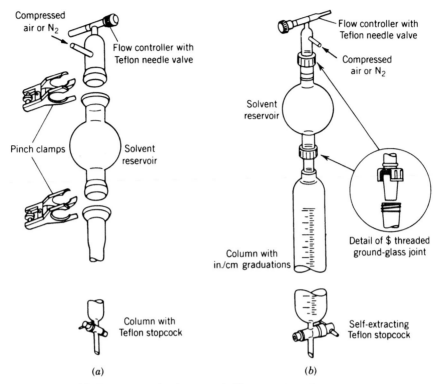

Figure 5.31 (a) Conventional column and (b) screw-threaded column.

to collect the fractions. This technique has been used successfully to separate mixtures ranging from 150 to 1000 mg.

Thin-layer chromatography

Thin-layer chromatography (TLC) is closely related to column chromatography, in that the phases used in both techniques are essentially identical. Alumina and silica gel are typical stationary phases, and the usual solvents are the mobile phases. There are, however, some distinct differences between TLC and column chromatography. The mobile (liquid) phase *descends* in column chromatography; the mobile phase *ascends* in TLC. The column of stationary-phase material used in column chromatography is replaced by a thin layer (100 μm of stationary phase spread over a flat surface. A piece of window glass, a microscope slide, or a sheet of plastic can be used as the support for the thin layer of stationary phase. It is possible to prepare your own glass plates, but plastic-backed thin-layer plates are only commercially available. Plastic-backed plates are particularly attractive because they can easily be cut with scissors into strips of any size. Typical strips measure about 1 $\times$ 3 in., but even smaller strips can be satisfactory.

Thin-layer chromatography has some distinct advantages: it needs little time (2–5 min), and it needs *very* small quantities of material (2–20 μg). The chief disadvantage of this type of chromatography is that it is not very amenable to preparative scale work. Even when large surfaces and thicker layers are used, separations are most often restricted to a few miligrams of material.

Note _____

> Do not touch the active surface of the plates with your fingers. Handle them only by the edge.

TLC is performed as follows:

Step 1. Draw a light pencil line parallel to the short side of the plate, 5–10 mm from the edge. Mark one or two points, evenly spaced, on the line. Place the sample to be analyzed (1 mg or less) in a 100-μL conical vial and add a few drops of a solvent to dissolve the sample. Use a capillary micropipet to apply a small fraction of the solution from the vial to the plate (Fig. 5.32). (These pipets are prepared by the same technique used for constructing the capillary insert for ultramicro boiling-point determinations, see Chapter 4.) Apply the sample to the adsorbent side of the TLC plate by gently touching the tip of the filled capillary to the plate. Remove the tip from the plate before the dot of solvent grows to much more than a few millimeters in diameter. If it turns out that you need to apply more sample, let the dot of solvent evaporate and then reapply more sample to exactly the same spot.

Step 2. Place the spotted thin-layer plate in a screw-capped, wide-mouth jar, or a beaker with a watch glass cover, containing a small amount of elution solvent (Fig. 5.33). It helps if one side of the jar's (beaker's) interior is covered with a piece of filter paper that wicks the solvent up to increase the surface area of the solvent. The TLC plate must be positioned so that the spot of your sample is *above* the solvent. Cap the jar, or replace the watch glass on the beaker, to maintain an atmosphere saturated with the elution solvent. The elution solvent climbs the plate by capillary action, eluting the sample up the plate. *Do not move the developing chamber after the action has started.* Separation of mixtures into individual spots occurs by exactly the same mechanism as in column chromatography. Stop the elution by removing the plate from the jar or beaker when the solvent front nears the top of the TLC plate. Quickly (before the solvent evaporates) mark the position of the solvent front on the plate.

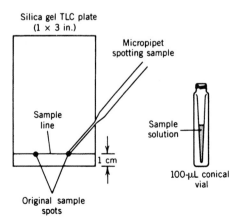

Figure 5.32 Applying a sample to a TLC plate.

Figure 5.33 Developing a TLC plate. *From Zubrick J. W. The Organic Chem Lab Survival Manual, 4th ed.; Wiley: New York. 1997. Reprinted by permission of John Wiley & Sons, Inc., New York.)*

Step 3. Colorless, separated components of a mixture can often be observed in a developed TLC plate by placing the plate in an iodine-vapor chamber (a sealed jar containing solid I_2) for a minute or two. Iodine forms a reversible complex with most organic substances and dark spots will thus appear in those areas containing sample material. Mark the spots with a pencil soon after removing the TLC plate from the iodine chamber because the spots may fade. Samples that contain a UV-active chromophore (see Chapter 8) can be observed without using iodine. TLC plates are commonly prepared with an UV-activated fluorescent indicator mixed in with the silica gel. Sample spots can be detected with a hand-held UV lamp; the sample quenches the fluorescence induced by the lamp and appears as a dark spot against the fluorescent blue-green background.

Step 4. The TLC properties of a compound are reported as R_f values (retention factors). The R_f value is the distance traveled by the substance divided by the distance traveled by the solvent front (this is why the position of the solvent front should be quickly marked on the plate when the chromatogram is terminated; see Fig. 5.34). TLC R_f values vary with the moisture content of the adsorbent. Thus, the actual R_f of a compound in a given solvent can vary from day to day and from laboratory to laboratory. The best way to determine if two samples have identical R_f values is to elute them together on the same plate.

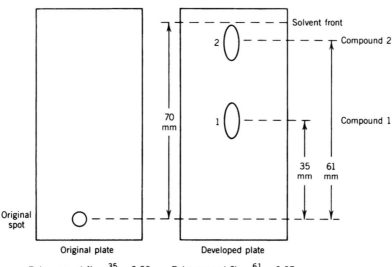

$$R_f\text{(compound 1)} = \frac{35}{70} = 0.50 \qquad R_f\text{(compound 2)} = \frac{61}{70} = 0.87$$

Figure 5.34 Determining R_f values.

Thin-layer chromatography is used in a number of applications. The speed of the technique makes it quite useful for monitoring large-scale column chromatography. Analysis of fractions can guide decisions on the solvent-elution sequence. TLC analysis of column-derived fractions can also determine how best to combine collected fractions. Following the progress of a reaction by periodically removing small aliquots for TLC analysis is an extremely useful application of thin-layer chromatography.

Paper chromatography

The use of cellulose paper as an adsorbent is referred to as *paper chromatography*. This technique has many of the characteristics of TLC in that sheets or strips of filter paper are used as the stationary phase. In this case, however, the paper is usually positioned to hang down from trays holding the paper and the elution solvent. The solvent front, therefore, descends downward rather than upward as in TLC. Paper chromatography has a distinct advantage: It is very amenable to the use of aqueous mobile phases and very small sample sizes. It is primarily used for the separation of highly polar or polyfunctional species such as sugars and amino acids. It has one major disadvantage: It is very slow. Paper chromatograms can easily take three to four hours or more to elute.

High-performance liquid chromatography

Although gas chromatography is a powerful chromatographic method, it is limited to compounds that have a significant vapor pressure at temperatures up to about 200 °C. Thus, compounds of high molecular weight and/or high polarity cannot be separated by GC. High-performance liquid chromatography (HPLC) does not present this limitation.

GC and HPLC are somewhat similar, in the instrumental sense, in that the analyte is partitioned between a stationary phase and a mobile phase. Whereas the mobile phase in GC is a gas, the mobile phase in HPLC is a liquid. As shown schematically in Figure 5.35, the mobile phase (solvent) is delivered to the system by a pump capable of pressures up to about 6000 psi. The sample is introduced by the injection of a solution into an injection loop. The injection loop is brought in line between the pump and the column (stainless steel) by turning a valve; the sample then flows down the column, is partitioned, and flows out through a detector.

The solid phase in HPLC columns used for organic monomers is usually some form of silica gel. "Normal" HPLC refers to chromatography using a solid phase (usually silica gel) that is more polar than the liquid phase, or solvent, so that the less polar compounds elute more rapidly. Typical solvents include ethyl acetate, hexane, acetone, low molecular weight alcohols, chloroform, and acetonitrile. For extremely polar compounds, such as amino acids, "reversed-phase" HPLC is used. Here, the liquid phase is more polar than the stationary phase, and the more polar compounds elute more rapidly. The mobile phase is usually a mixture of water and a water-miscible organic solvent such as acetonitrile, dioxane, methanol, isopropanol, or acetone. The stationary phase is

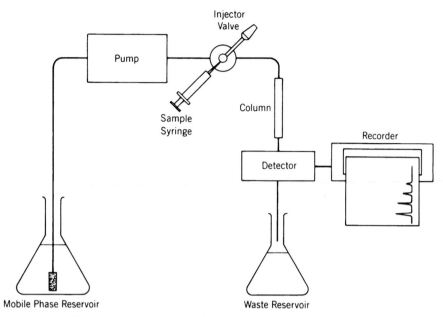

Figure 5.35 High-performance liquid chromatography system block diagram. *(Courtesy of the Perkin-Elmer Corp., Norwalk, CT.)*

usually a modified silica gel where the —OH groups of the silica gel have been replaced by —OSiR$_3$ groups; R is typically a linear C$_{18}$ alkyl chain. These so-called "bonded-phase" columns are not capable of handling as much analyte as normal silica gel columns, and are thus easily overloaded and are less useful for preparative work. (For further discussion see Solid-Phase Extraction, Technique 4.)

A wide variety of detection systems are available for HPLC. UV detection is common, inexpensive, and sensitive. The solvent flowing off the column is sent through a small cell where the UV absorbance is recorded over time. Many detectors are capable of variable wavelength operation so the detector can be set to the wavelength most suitable to the compound or compounds being analyzed. Photodiode array detectors are available; these can obtain a full UV spectrum in a fraction of a second, so that more information can be obtained on each component of a mixture. For compounds that absorb light in the visible (vis) spectrum, many detectors can be set to visible wavelengths. The principal shortcoming of UV-vis detection is that to be detected, compounds being studied must have a UV chromophore, such as an aromatic ring or other conjugated π system (see Chapter 8).

For compounds that lack a UV-vis chromophore, refractive index (RI) detection is a common substitute. An RI detector measures the difference in refractive index between the eluant and a reference cell filled with the elution solvent. Refractive index detection is significantly less sensitive than UV-vis detection, and the detector is quite sensitive to temperature changes during the chromatographic run.

More sophisticated HPLC instruments offer the ability to mix two or three different solvents and to use solvent gradients by changing the solvent composition as the chromatographic run progresses. This allows the simultaneous analysis of compounds that differ greatly in their polarity. For example, a silica gel column might begin elution with a very nonpolar solvent, such as hexane. The solvent polarity is then continuously increased by blending in more and more ethyl acetate until the elution solvent is pure ethyl acetate. This effect is directly analogous to temperature programming in GC.

For analytical work, typical HPLC columns are about 5 mm in diameter and about 25 cm in length. The maximum amount of analyte for such columns is generally less than 1 mg, and the minimum amount is determined by the detection system. High-performance liquid chromatography can thus be used to obtain small amounts of purified compounds for infrared (IR), nuclear magnetic resonance (NMR), or mass spectrometric (MS) analysis. Larger "semi-preparative" columns that can handle up to about 20 mg without significant overloading are useful for obtaining material for ^{13}C NMR spectroscopy or further synthetic work.

HPLC has the advantage that it is rapid, it uses relatively small amounts of solvent, and it can accomplish very difficult separations.

TECHNIQUE 6B CONCENTRATION OF SOLUTIONS

The solvent can be removed from chromatographic fractions (extraction solutions, or solutions in general) by a number of different methods.

Distillation

Concentration of solvent by distillation is straightforward. This approach allows for high recovery of volatile solvents and often can be done outside a hood. The Hickman still head and the 5- or 10-mL round-bottom flask are useful for this purpose. Distillation should be used primarily for concentration of the chromatographic fraction, followed by transfer of the concentrate with a Pasteur filter pipet to a vial for final isolation.

Evaporation with nitrogen gas

A very convenient method for removal of final solvent traces is the concentration of the last 0.5 mL of a solution by evaporation with a gentle stream of nitrogen gas while the sample is warmed in a sand bath. This process is usually done at a hood station where several Pasteur pipets can be attached to a manifold leading to a source of dry nitrogen gas. Gas flow to the individual pipets is controlled by needle valves. *Always test the gas flow with a blank vial of solvent.*

Ruekberg described an alternative way to remove solvent from solutions of compounds that are not readily oxidized.[24] The setup includes an aquarium air pump, a pressure safety valve and ballast container, a drying tube, and a manifold. Blunted hypodermic needles are used in place of Pasteur pipets.

---

[24]Ruekberg, B. *J. Chem. Educ.* **1995**, *72*, A200.

The sample vial will cool as the solvent evaporates, and gentle warming and agitation of the vial will thus help remove the last traces of the solvent. This avoids possible moisture condensation on the sample residue, as long as the gas itself is dry. *Do not leave the heated vial in the gas flow after the solvent is removed!* This precaution is particularly important in the isolation of liquids. Tare the vial before filling it with the solution to be concentrated; constant weight over time is the best indication that all solvent has been removed.

Removal of solvent under reduced pressure

Concentration of solvent under reduced pressure is very efficient. It reduces the time for solvent removal in microscale experiments to a few minutes. In contrast, distillation or evaporation procedures require many minutes for even relatively small volumes. Several methods are available.

FILTER FLASK METHOD

This vacuum-concentration technique can be tricky and should be practiced prior to committing hard-won reaction product to this test. The procedure is most useful with fairly large chromatographic fractions (5–10 mL). The sequence of operations is as follows (see also Fig. 5.36):

Step 1. Transfer the chromatographic fraction to the 25-mL filter flask.

Step 2. Insert the 11-mm Hirsch funnel and rubber adapter into the flask.

Step 3. Turn on the water pump (with trap) and connect the vacuum tubing to the pressure flask side arm while holding the flask in one hand.

Step 4. Place the thumb of the hand holding the filter flask over the Hirsch funnel filter bed to shut off the air flow through the system (see Fig. 5.36). This will result in an immediate drop in pressure. The volatile solvent will rapidly come to a boil at room temperature. Thumb pressure adjusts air leakage through the Hirsch funnel and thereby controls the pressure in the system. It is also good practice to learn to manipulate the pressure so that the liquid does not foam up into the side arm of the filter flask.

The filter flask must be warmed by the sand bath during this operation; rapid evaporation of the solvent will quickly cool the solution. The air leak used to control the pressure results in a stream of moist laboratory air being rapidly drawn over the surface of the solution. If the evaporating liquid becomes cold, water will condense over the interior of the filter flask and contaminate the isolated residue. Warming the flask while evaporating the solvent

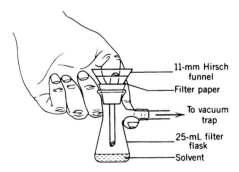

11-mm Hirsch funnel
Filter paper
To vacuum trap
25-mL filter flask
Solvent

Figure 5.36 Removal of solvent under reduced pressure.

will avoid this problem and help speed solvent removal. The temperature of the flask should be checked from time to time by touching it with the palm of the free hand. The flask is kept slightly above room temperature by adjusting the heating and evaporation rates. It is best to practice this operation a few times with pure solvent (blanks) to see whether you can avoid boilovers and accumulating water residue in the flask.

ROTARY EVAPORATOR METHOD

In most research laboratories, the most efficient way to concentrate a solution under reduced pressure is to use a **rotary evaporator.** A commercial micro-rotary evaporator is shown in Figure 5.37. This equipment makes it possible to recover the solvent removed during the operation.

The rotary evaporator is a motor-driven device that rotates the flask containing the solution to be concentrated under reduced pressure. The rotation continuously exposes a thin film of the solution for evaporation. This process is very rapid, even well below the boiling point of the solvent being removed. Since the walls of the rotating flask are constantly rewetted by the solution, bumping and superheating are minimized. The rotating flask may be warmed in a water bath or other suitable device that controls the rate of evaporation. A suitable adapter (a "bump bulb") should be used on the rotary evaporator to guard against splashing and sudden boiling, which may lead to lost or contaminated products.

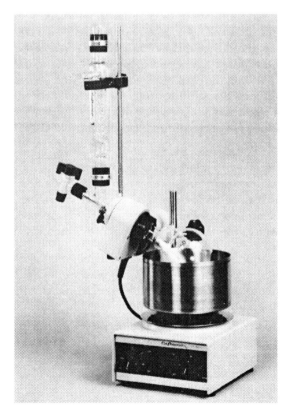

Figure 5.37 Heidolph micro-rotary evaporator. (*Courtesy of Caframo, Ltd., Wiarton, Ontario, Canada.*)

In microscale work, never pour a recovered liquid product from the rotary flask. Always use a Pasteur pipet.

HICKMAN STILL–ROTARY EVAPORATION APPARATUS

A simple microscale rotary evaporator for use in the instructional laboratory consists of a 10-mL round-bottom flask connected to a capped Hickman still (sidearm type), which in turn is attached to a water aspirator (with trap).[25] The procedure involves transferring the solution to be concentrated to the preweighed 10-mL flask. The flask is then attached to a Hickman still with its top joint sealed with a rubber septum and threaded compression cap. The apparatus is connected by the still sidearm to the trap–vacuum source with a vacuum hose. With the aspirator on, one shakes the apparatus while warming the flask in the palm of the hand. In this manner, bumping is avoided and evaporation is expedited. The still acts as a splash guard. Heat transfer is very effective, and once the flask reaches ambient temperature, the vacuum is released by venting through the trap stopcock.

QUESTIONS

5-42. When marking the sample line on a TLC plate, why is it inadvisable to use a ball-point pen?

5-43. A series of dyes is separated by TLC. The data are given below. Calculate the R_f value for each dye.

MATERIAL	DISTANCE MOVED (cm)
Solvent	6.6
Bismarck brown	1.6
Lanacyl violet BF	3.8
Palisade yellow 3G	5.6
Alizarine emerald G	0.2

5-44. Why is it important not to let the level of the elution solvent in a packed chromatographic column drop below the top of the solid-phase adsorbent?

5-45. What are some advantages of using column chromatography to purify reaction products in the microscale laboratory?

5-46. Discuss the similarities and dissimilarities of TLC, paper, and column chromatography.

5-47. Discuss the similarities and dissimilarities of HPLC and gas chromatography.

5-48. **(a)** What are the main advantages of using flash chromatography?
(b) How can TLC be used in connection with flash chromatography?

---

[25]Maynard, D. F. *J. Chem. Educ.* **1994,** *71,* A272.

TECHNIQUE 7
Collection or Control of Gaseous Products

Water-insoluble gases

Numerous organic reactions lead to the formation of gaseous products. If the gas is insoluble in water, collection is easily accomplished by displacing water from a collection tube. A typical experimental setup for the collection of gases is shown in Figure 5.38.

As illustrated, the glass capillary efficiently transfers the evolved gas to the collection tube. The delivery system need not be glass; small polyethylene or polypropylene tubing may also serve this purpose. In this arrangement, a syringe needle is inserted through a septum to accommodate the plastic tubing as shown in Figure 5.39. An alternative to this connector is a shortened Pasteur pipet inserted through a thermometer adapter (Fig. 5.39). Another alternative to the syringe needle or glass pipet tip is suggested by Jacob.[26] The lower half of the tapered tip of a plastic automatic delivery pipet tip is cut off to prevent buildup of excess pressure in the reaction vessel. The pipet tip is then inserted through a previously pierced rubber septum or into a thermometer adapter. The narrow end of the tip is then inserted into the plastic tubing.

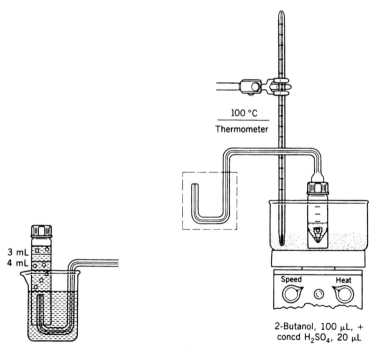

Figure 5.38 Microscale gas collection apparatus.

---

[26]Jacob, L. A. *J. Chem. Educ.* **1992,** *69*, A313.

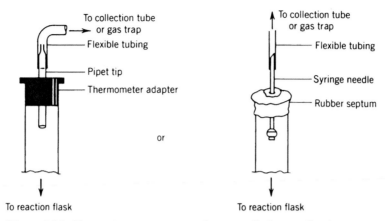

Figure 5.39 Alternative arrangements for controlled gas collection.

An example of a reaction leading to gaseous products that can use this collection technique is the acid-catalyzed dehydration of 2-butanol. The products of this reaction are a mixture of alkenes: 1-butene, *trans*-2-butene and *cis*-2-butene, which boil at -6.3, 0.9, and 3.7 °C, respectively.

In Figure 5.38 the gas collection tube is capped with a rubber septum. This arrangement allows for convenient removal of the collected gaseous butenes using a gas-tight syringe, as shown in Figure 5.40. In this particular reaction, the mixture of gaseous products is conveniently analyzed at ambient temperature by GC (see Technique 1).

Trapping byproduct gases

Some organic reactions release poisonous or irritating gases as byproducts. For example, hydrogen chloride, ammonia, and sulfur dioxide are typical byproducts in organic reactions. In these cases, the reaction is generally run in a **hood.** A gas trap may be used to prevent the gases from being released into the

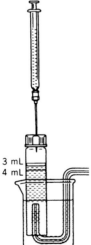

Figure 5.40 Removal of collected gases.

laboratory atmosphere. If the evolved gas is water soluble, the trap technique works well at the microscale level. The evolved gas is directed from the reaction vessel to a container of water or other aqueous solution, wherein it dissolves (reacts). For example, a dilute solution of sodium or ammonium hydroxide is suitable for acidic gases (such as HCl); a dilute solution of sulfuric or hydrochloric acid is suitable for basic gases (such as NH_3 or low molecular weight amines). Various designs are available for gas traps. A simple, easily assembled one for a gas that is very soluble in water is shown in Figure 5.41. Note that the funnel is not immersed in the water. If the funnel is held below the surface of the water and a large quantity of gas is absorbed or dissolved, the water easily could be drawn back into the reaction assembly. If the gas to be collected is not very soluble, the funnel may be immersed just below (1–2 mm) the surface of the water.

At the microscale level, small volumes of gases are evolved that may not require the funnel. Three alternatives are available:

a. Fill the beaker (100 mL) in Figure 5.41 with moistened fine glass wool and lead the gas delivery tube directly into the wool.

b. Place moistened glass wool in a drying tube and attach the tube to the reaction apparatus (see ⬤ , Fig. 3.10W). However, be careful not to let the added moisture drip into the reaction vessel; place a small section of **dry** glass wool in the tube before the moist section is added.

c. Use a water aspirator. An inverted funnel can be placed over the apparatus opening where the evolved gas is escaping (usually the top of a condenser) and connected with flexible tubing (through a water trap) to the aspirator. A second arrangement is to use a glass or plastic T-tube (open on one end) inserted in the top of the condenser, by use of a rubber stopper, in place of the funnel.[27] If the reaction must be run under anhydrous conditions, a drying tube is inserted between the condenser and T-tube. This arrangement is very efficient, easy to assemble, and inexpensive. The simplest method is to clamp a Pasteur pipet so that its tip is inserted well into the condenser, and connect it (through a water trap) to the aspirator.

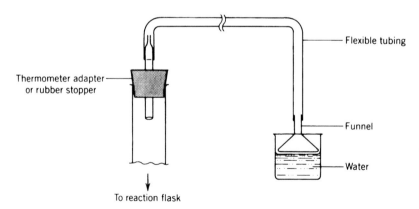

Figure 5.41 Trapping a water-soluble gas.

[27]Horodniak, J. W.; Indicator, N. *J. Chem. Educ.* **1970**, *47*, 568.

QUESTIONS

5-49. In Figure 5.38 why is a septum, not just a plain cap, used on the top of the gas collection tube?

5-50. An evolved gas is directed from the reaction vessel to a container of water or other aqueous solution, wherein it dissolves (reacts). For example, a dilute solution of sodium or ammonium hydroxide is suitable for acidic gases (such as HCl). What solution would be appropriate to trap thiols and sulfides? (*Hint:* Consult a qualitative analysis text.)

5-51. One way to eliminate emissions is to place moistened glass wool in a drying tube, which is then attached to the reaction apparatus (, Fig. 3.10W). What precautions must be taken when using this method?

5-52. In the collection of water-insoluble gases with the apparatus shown in Figure 5.38, describe how one might measure the rate at which a gas is evolved from a reaction mixture.

TECHNIQUE 8

Measurement of Specific Rotation

Solutions of optically active substances, when placed in the path of a beam of polarized light, may rotate the plane of the polarized light. Enantiomers (two molecules that are nonidentical mirror images) have identical physical properties (melting points, boiling points, infrared and nuclear magnetic resonance spectra, etc.) except for their interaction with plane polarized light, their *optical activity*. Optical rotation data can provide important information concerning the absolute configuration and the enantiomeric purity of a sample.

Optical rotation is measured using a *polarimeter*. This technique is applicable to a wide range of analytical problems, from purity control to the analysis of natural and synthetic compounds. The results obtained from measuring the observed angle of rotation α are generally expressed as the *specific rotation* $[\alpha]$.

Theory

Ordinary light behaves as though it were composed of electromagnetic waves in which the oscillating electric field vectors are distributed among the infinite number of possible orientations around the direction of propagation (see Fig. 5.42).

Figure 5.42 Oscillation of the electric field of ordinary light occurs in all possible planes perpendicular to the direction of propagation. *(From Solomons, T.W.G* Organic Chemistry, *5th ed.; Wiley: New York, 1992. Reprinted by permission of John Wiley & Sons, Inc., New York.)*

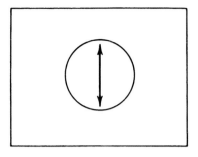

Figure 5.43 The plane of oscillation of the electric field of plane-polarized light. In this example the plane of polarization is vertical. *(From Solomons, T.W.G. Organic Chemistry, 5th ed.; Wiley: New York, 1992. Reprinted by permission of John Wiley & Sons, Inc., New York.)*

Note

A beam of light behaves as though it is composed of two, mutually perpendicular, oscillating fields: an electric field and a magnetic field. The oscillating magnetic field is not considered in the following discussion.

The planes in which the electrical fields oscillate are perpendicular to the direction of propagation of the light beam. If one separates one particular plane of oscillation from all other planes by passing the beam of light through a polarizer, the resulting radiation is plane-polarized (Fig. 5.43). In the interaction of light with matter, this plane-polarized radiation is represented as the vector sum of two circularly polarized waves. The electric vector of one of the waves moves in a clockwise direction; the other moves in a counterclockwise direction. Both waves have the same amplitude (Fig. 5.44). These two components add vectorially to produce plane-polarized light.

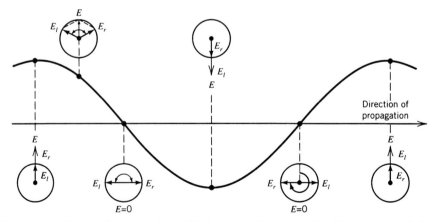

Figure 5.44 A beam of plane-polarized light viewed from the side (sine wave) and along the direction of propagation at specific times (circles) where the resultant vector E and the circularly polarized components E_l and E_r are shown. *(From Douglas, B., McDaniel, D. H., and Alexander, J. J. Concepts and Models of Inorganic Chemistry, 2d ed. Wiley, New York, 1983. Reprinted by permission of John Wiley & Sons, Inc. New York.)*

If the passage of plane-polarized light through a material reduces the velocity of one of the circularly polarized components more than the other by interaction with bonding and nonbonding electrons, the transmitted beam of radiation has its plane of polarization rotated from its *original* position (Figs. 5.45 and 5.46). A **polarimeter** is used to measure this angle of rotation.

The Polarimeter

The polarimeter measures the amount of rotation caused by an optically active compound (in solution) placed in the beam of the plane polarized light. The principal parts of the instrument are diagrammed in Figure 5.45. Two Nicol prisms are used in the instrument. The first prism, which polarizes the original light source, is called the *polarizer*. The second prism, called the *analyzer*, is used to examine the polarized light after it passes through the solution being analyzed.

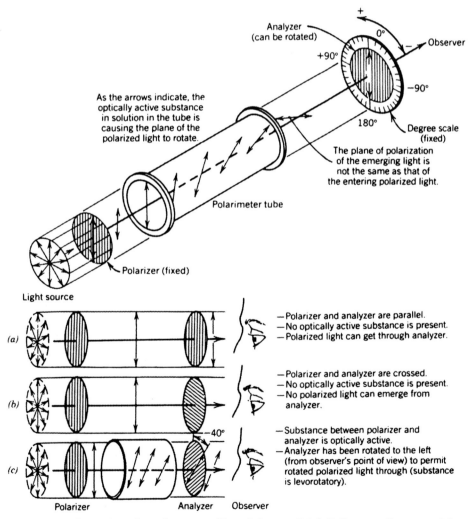

Figure 5.45 Operation of a polarimeter. (*From Solomons, T. W. G. Organic Chemistry, 5th ed. Wiley, New York, 1992, Reprinted by permission of John Wiley & Sons, Inc. New York.*)

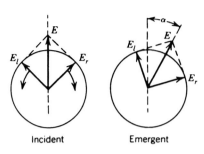

Figure 5.46 Plane-polarized light before entering and after emerging from an optically active substance. *(From Douglas, B., McDaniel, D. H., and Alexander, J. J. Concepts and Models of Inorganic Chemistry, 2d ed. Wiley, New York, 1983. (Reprinted by permission of John Wiley & Sons, Inc., New York.)*

When the axes of the analyzer and polarizer prisms are parallel (0°) and no optically active substance is present, the maximum amount of light is transmitted. If the axes of the analyzer and polarizer are at right angles to each other (90°), no transmission of light is observed. Placing an optically active solution into the path of the plane-polarized light causes one of the circularly polarized components to be slowed more than the other. The refractive indices are, therefore, different in the two circularly polarized beams. Figure 5.45 represents a case in which the left-hand component has been affected the most.

Note

In the simplified drawing, Figure 5.45, the effect on only one of the circularly polarized waves is diagrammed. See Figure 5.46 for a more accurate description (view from behind the figure).

This tilts the plane of polarization. The analyzer prism must be rotated to the left to maximize the transmission of light. If rotation is counterclockwise, the angle of rotation is defined as (−) and the enantiomer that caused the effect is called levorotatory (*l*). Conversely, clockwise rotation is defined as (+), and the enantiomer is dextrorotatory (*d*). Tilting the plane of polarization is called *optical activity*. Note that if a solution of equal amounts of a *d* and an *l* enantiomeric pair is placed in the beam of the polarimeter, no rotation is observed. Such a solution is *racemic*; it is an equimolar mixture of enantiomers.

The magnitude of optical rotation depends on several factors: (1) the nature of the substance, (2) the path length through which the light passes, (3) the wavelength of light used as a source, (4) the temperature, (5) the concentration of the solution used to make the measurement of optical activity, and (6) the solvent used in making the measurement.

The results obtained from the measurement of the observed angle of rotation, α_{obs}, are generally expressed in terms of *specific rotation* [α]. The sign and magnitude of [α] are dependent on the specific molecule and are determined by complex features of molecular structure and conformation; they cannot be easily explained or predicted. The specific rotation is a physical constant characteristic of a substance. The relationship of [α] to α_{obs} is as follows:

$$[\alpha]_\lambda^T = \frac{\alpha_{obs}}{lc}$$

where T = temperature of the sample in degrees Celsius (°C)

l = the length of the polarimeter cell in decimeters (1 dm = 0.1 m = 10 cm)

c = the concentration of the sample in grams per milliliter (g/mL),

λ = the wavelength of the light in nanometers (nm) used in the polarimeter.

These units are traditional, though most are esoteric by contemporary standards. The specific rotation for a given compound depends on both the concentration and the solvent, and thus both the solvent and concentration used must be specified. For example, $[\alpha]_D^{25}$ (c = 0.4, $CHCl_3$) = 12.3° implies that the measurement was recorded in a $CHCl_3$ solution of 0.4 g/mL at 25 °C using the sodium D line (589 nm) as the light source.

For increased sensitivity, many simple polarimeters have an optical device that divides the viewed field into three adjacent parts (triple-shadow polarimeter; Fig. 5.47). A very slight rotation of the analyzer will cause one portion to become dimmer and the other lighter (Fig. 5.47*a* and 5.47*c*). The angle of rotation reading (α) is recorded when the field sections all have the same intensity. An accuracy of ± 0.1° can be obtained.

INACCURATE MEASUREMENTS.

Several conditions may lead to inaccurate measurements, including trapped air bubbles in the cell, and solid particles suspended in the solution. Filter the solution, if necessary.

HIGH-PERFORMANCE POLARIMETERS AND OPTICAL ROTARY DISPERSION

For details of these two related topics refer to 🌀, Technique 8.

APPLICATIONS TO STRUCTURE DETERMINATION IN NATURAL PRODUCTS

Natural products provide interesting opportunities for measuring optical activity. An excellent example is the lichen metabolite, usnic acid, which can be easily isolated from its native source, "Old Man's Beard" lichens, as golden crystals.

Usnic acid

Usnic acid contains a single stereocenter (stereogenic center, or chiral center) and, therefore, has the possibility of existing as an enantiomeric pair of stereoisomers. Generally, in a given lichen, only one of the stereoisomers (R or S) is present. Usnic acid has a very high specific rotation (~ ± 460°), which makes it an ideal candidate for optical rotation measurements at the microscale level.

(a)   (b)   (c)

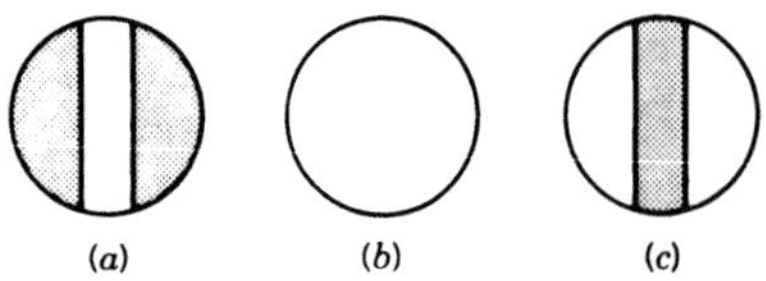

Figure 5.47 View through the eyepiece of the polarimeter. The analyzer should be set so that the intensity in all parts of the field is the same (b). When the analyzer is displaced to one side or the other, the field will appear as in (a) or (c).

QUESTIONS

5-53. A solution of 300 mg of optically active 2-butanol in 10 mL of water shows an optical rotation of $-0.54°$. What is the specific rotation of this molecule?

5-54. Draw the structure of usnic acid and locate its stereocenter.

5-55. If a solution of an equimolar mixture of an enantiomeric pair is placed in the beam path of the polarimeter, what would you observe?

5-56. The specific rotation of $(+)Q$ is $+12.80°$. At identical concentration, solvent, pathlength, and light wavelength, the observed rotation of a solution containing both enantiomers of Q is $6.40°$. What are the relative concentrations of each enantiomer in the solution?

TECHNIQUE 9
Sublimation

Sublimation is especially suitable for purifying solids at the microscale level. It is useful when the impurities present in the sample are nonvolatile under the conditions used. Sublimation is a relatively straightforward method; the impure solid need only be heated.

Sublimation has additional advantages: (1) It can be the technique of choice for purifying heat-sensitive materials—under high vacuum it can be effective at low temperatures; (2) solvents are not involved and, indeed, final traces of solvents are effectively removed; (3) impurities most likely to be separated are those with lower vapor pressures than the desired substance and often, therefore, lower solubilities, exactly those materials very likely to be contaminants in a recrystallization; (4) solvated materials tend to desolvate during the process; and (5) in the specific case of water of solvation, it is very effective even with substances that are deliquescent. The main disadvantage of the technique is that it can be less selective than recrystallization when the vapor pressure of the desired material being sublimed is similar to that of an impurity.

Some materials sublime at atmospheric pressure (CO_2, or dry ice, is a well-known example), but most sublime when heated below their melting points under reduced pressure. The lower the pressure, the lower the sublimation temperature. Substances that do not have strong intermolecular attractive forces are excellent candidates for purification by sublimation. Napthalene, ferrocene, and *p*-dichlorobenzene are examples of compounds that are readily sublimed.

Sublimation theory

Sublimation and distillation are closely related. Crystals of solid substances that sublime, when placed in an evacuated container, will gradually generate molecules in the vapor phase by the process of evaporation (i.e., the solid has a vapor pressure). Occasionally, one of the vapor molecules will strike the crystal surface or the walls of the container and be held by attractive forces. This process, condensation, is the reverse of evaporation.

Sublimation is the complete process of evaporation from the solid phase to condensation from the gas phase to directly form crystals without passing through the liquid phase.

A typical single-component phase diagram is shown in Figure 5.48, which relates the solid, liquid, and vapor phases of a substance to temperature and pressure. Where two of the areas (solid, liquid, or vapor) touch, there is a line, and along each line the two phases exist in *equilibrium*. Line *BO* is the sublimation–vapor pressure curve of the substance in question; only along line *BO* can solid and vapor exist together in equilibrium. At temperatures and pressures along the *BO* curve, the liquid state is thermodynamically unstable. Where the three lines representing pairs of phases intersect, all three phases exist together in equilibrium. This point is called the *triple point*.

Many solid substances have a sufficiently high vapor pressure near their melting point that allows them to be sublimed easily under reduced pressure in the laboratory. Sublimation occurs when the vapor pressure of the solid equals the pressure of the sample's environment.

Experimental setup

Heating the sample with a microburner or a sand bath to just below the melting point of the solid causes sublimation to occur. Vapors condense on the cold-finger surface, whereas any less volatile residue will remain at the bottom of the flask. Apparatus for sublimation of small quantities are now commercially available (Fig. 5.49). Two examples of simple, inexpensive apparatus suitable for sublimation of small quantities of material in the microscale organic laboratory are shown in Figure 5.50.

An example of the purification of a natural product, where the sublimation technique at the microscale level is effective, is the case of the alkaloid caffeine. This substance can be isolated by extraction from tea.

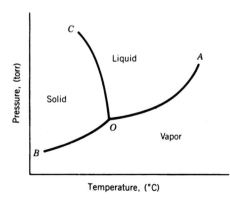

Figure 5.48 Single-component phase diagram.

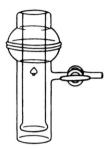

Figure 5.49 Vacuum sublimator. *(Courtesy of ACE Glass Inc., Vineland, NJ.)*

Precautions

Several precautions should be observed when performing a sublimation:

1. If you use the first setup in Figure 5.50, make sure you attach the hose connections to the cold finger in the proper manner. *The incoming cold water line is attached to the center tube.*

2. If you generate a vacuum using a water aspirator, make sure you place a water trap in the line. *Apply the vacuum to the system before you turn on the cooling water to the condenser.* This will keep moisture in the air in the flask from being condensed on the cold finger. Let the cold finger warm up before releasing the vacuum.

3. After the sublimation is complete, release the vacuum *slowly* so as not to disturb the sublimed material.

4. When using either of the arrangements in Figure 5.50, be careful to avoid loss of purified product as you remove the cold finger from the assembly.

5. The distance between the tip of the cold finger and the bottom of the sublimator should be less than 1 cm in most cases.

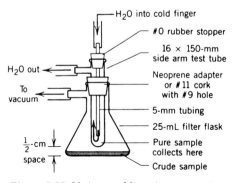
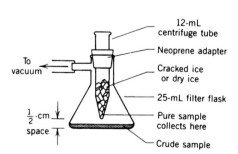

Figure 5.50 Various sublimation apparatus.

QUESTIONS

5-57. List the advantages and disadvantages of sublimation as a purification technique.

5-58. For a solid compound to evaporate at atmospheric pressure it must have an unusually high vapor pressure. What molecular structural features contribute to this vapor pressure?

5-59. Why apply the vacuum to the sublimation system *before* you turn on the cooling water to the water condenser?

5-60. Why place a water trap in the vacuum line when using an aspirator to obtain the vacuum?

5-61. Why is sublimation particularly useful for purifying deliquescent compounds?

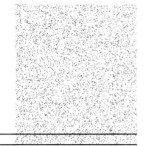

6

Introduction to
Infrared Spectroscopy

TECHNIQUE 10

Introduction

The wavelike character of electromagnetic radiation can be expressed in terms of velocity v, frequency ν, and wavelength λ of sinusoidally oscillating electric and magnetic vectors traveling through space (Fig. 6.1). Frequency is defined as the number of waves passing a reference point per unit time, usually expressed as cycles per second (s^{-1}) or hertz (Hz). The velocity of the wave, therefore, equals the product of frequency and wavelength:

$$v = \nu\lambda$$

If the wavelength (the distance between the wave maxima or alternate nodes) is measured in centimeters, v is expressed in centimeters per second (cm/s). For radiation traveling in a vacuum, v becomes a constant, c $(c \sim 3 \times$

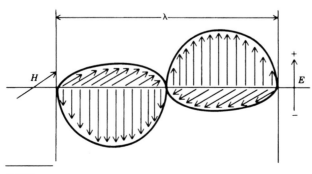

Figure 6.1
Electromagnetic wave. H, magnetic field; E, electric field; λ wavelength.

10^{10} cm/s), for all wavelengths. When electromagnetic radiation traverses other media, however, the velocity changes. The ratio of the speed in a vacuum, c, to the matrix velocity, v, is termed the *refractive index, n,* of the material:

$$n = \frac{c}{v}$$

Since n is frequency dependent, the frequency at which the refractive index is measured must be specified. Frequency, however, has been shown to be independent of the medium and, therefore, remains constant. Wavelength thus varies inversely with n.

$$\lambda = \frac{c}{nv}$$

Since the velocity of electromagnetic radiation in a vacuum is normally greater than that in any other medium, n will generally be greater than 1 at all frequencies. Thus, the wavelength must become shorter for a particular frequency when measured in any matrix.

Frequency can be considered to be a more fundamental property of radiation because it is independent of the medium. This property also requires that the energy E associated with the radiation be matrix independent because E is directly proportional to frequency by

$$E = h\nu$$

where E equals the energy of a photon, which is related to frequency ν by Planck's constant (h) (6.6×10^{-27} erg s or 6.6×10^{-34} J s).

The vibrational states present in molecules can be excited by absorption of photons. The nuclear masses and bond force constants determine the separation of these states and, therefore, the energies of the photons involved in the absorption process. The corresponding radiation frequencies fall predominantly in the IR region ($10^{14} - 10^{12}$ Hz) of the electromagnetic spectrum.

The IR spectrum is currently measured in *wavenumbers, $\tilde{\nu}$,* which are units proportional to frequency and energy. The wavenumber is defined as

$$\tilde{\nu} = \frac{\nu}{c} = \frac{E}{hc}$$

and as

$$\nu = \frac{c}{n\lambda} \qquad \text{then in air} \qquad \tilde{\nu} = \frac{\sim 1}{\lambda}$$

The wavenumber, as expressed in units of reciprocal centimeters (cm^{-1}) (the number of waves per centimeter), offers several advantages:

1. Wavenumbers are directly proportional to frequency and are expressed in much more convenient numbers (in this region of the spectrum), 5000–500 cm^{-1}.

2. As shown above, wavenumbers are easily converted to wavelength values. The reciprocal of $\tilde{\nu}$ and conversion of centimeters to wavelength units are all that is required (this is particularly handy because much early IR data were recorded linearly in wavelength). The wavelength unit employed in most of these spectra was the micron, μ (1×10^{-4} cm). The micron has been replaced by a unit expressed in meters, the micrometer, μm (1 μm $= 1 \times 10^{-6}$ m).

3. Because the wavenumber is directly proportional to frequency and energy, the use of wavenumbers allows spectra to be displayed linearly in energy, which is a distinct aid in sorting out related vibrational transitions. For an introductory discussion of vibrational energy see 🌀 Chapter 6, IR section, Part I.

INTRODUCTION TO GROUP FREQUENCIES: INTERPRETATION OF INFRARED SPECTRA

Studies of the vibrational spectra of thousands of molecules have revealed that many of the normal modes associated with particular atomic arrangements may be transferred from one molecule to another. A *normal mode* of vibration is one of the residual *fundamental* vibrations of a molecular system in which the atomic displacements are all related by simple harmonic motion to the overall total vibrational motion (or vibrational energy) of the molecule. There are $3N - 6$ (where N = the number of atoms) *normal modes* (or fundamental vibrations or vibrational degrees of freedom—all these terms are essentially synonymous) present in all nonlinear molecules. Linear molecules have $3N - 5$ normal modes—in this case there is one more normal mode of vibrational energy present because a rotational degree of freedom has been lost. Rotation around the molecular axis involves no energy because the atomic nuclei are assumed to be point sources of matter. (For an introductory discussion of vibrational energy See 🌀 Chapter 6, IR section, Part I B.)

Operating under selection rules these normal modes of vibration give rise to absorption bands in the infrared region of the spectrum (see, for example, the infrared spectrum of *n*-hexane, Fig. 6.2, p. 152). In the analysis these modes are often assigned numbers. For example, the 30 modes of benzene (where $N = 12$ in the $3N - 6$ expression) can be assigned 1 through 30 or the numbering can be done using any one of a number of different criteria. Subscripts a and b are often used to indicate doubly degenerate modes, that is, modes that have identical energies (and thus required to have the same frequency). One of the numbering systems for the benzene ring is used here when the aromatic ring stretching vibrations are identified (see Table 6.6 and 🌀 Chapter 6, IR section, Part II D, for a more detailed discussion of normal modes).

Many of these vibrational frequencies are associated with small groups of atoms that are essentially uncoupled from the rest of the molecule. The absorption bands that result from these modes, therefore, are characteristic of the small group of atoms regardless of the composition of other parts of the molecule. These vibrations are known as the *group frequencies*. Interpretation of infrared spectra of complex molecules based on group frequency assignments is an extremely powerful aid in the elucidation of molecular structure.

The following four factors make significant contributions to the development of a good (useful, reliable) group frequency from a molecular vibration:

1. The group has a large dipole-moment change during vibrational displacement. This change in moment is formally related to the efficiency of absorption of radiation during the molecular displacement by the expression

$I \alpha \, (\delta\mu/\delta Q)^2$ (where I = intensity, μ = electric dipole moment, and Q = the normal coordinate [a mathematical description of the vibration]). Thus, if $\delta\mu/\delta Q$ is large, there is a large absorption of infrared radiation which gives rise to very intense bands (the intensity is dependent on the square of the moment change at that vibrational frequency).

2. The presence of a large force constant, so that for many of these groups the stretching frequency occurs at high values above the fingerprint region.

3. The fundamental mode occurs in a frequency range that is reasonably narrow (little coupling), but sensitive enough to the local environmental to allow for considerable interpretation of the surrounding structure.

4. The range of frequencies is determined by a number of factors that are now well understood in terms of the mass, geometric, electronic, intramolecular, and intermolecular effects (for an introductory discussion of these effects see ⬤ Chapter 6, IR section, Part III A).

STRATEGIES FOR INTERPRETING INFRARED SPECTRA

1. Divide the spectrum into two parts at 1350 cm^{-1}.

2. Above 1350 cm^{-1}, absorption bands have a high probability of being good group frequencies. The interpretation is usually reliable and free from ambiguities. We can be much more confident of our assignments in this region even with rather weak bands.

3. Because of the reliability of the high-wavenumber region, we always begin the interpretation of a spectrum at this end.

4. Bands below 1350 cm^{-1} may be either group frequencies or fingerprint frequencies.

5. Below 1350 cm^{-1}, group frequencies are less easily assigned. In addition, even if a reliable group frequency occurs in this region, absorption at that frequency is not necessarily a result of that mode. That is, fingerprint bands can also randomly occur in the same location as reliable group frequency bands and the observer cannot usually distinguish which type of band is present.

6. To make more confident assignments below 1350 cm^{-1}, it is helpful to be able to associate a secondary property, such as band shape, with the particular mode. For example, it helps to know whether the band is very intense, broad, sharp, occurs as a characteristic doublet, gives the correct frequency shift on isotopic substitution, or the like.

7. A good rule to remember is that in the fingerprint region the **absence** of a band is more important than the presence of a band. If a band is absent, you can conclude with confidence that a reliable group frequency assigned to this region is absent and therefore the group must be absent from the sample. At the same time you also know that no interfering fingerprint bands occur in the region.

8. Before beginning the interpretation, note the sampling conditions and determine as much other information about the sample as possible—such as molecular weight, melting point, boiling point, color, odor, elemental analysis, solubility, and refractive index.

9. In the interpretation try to assign the most intense bands first. These bands very often will be associated with a polar functional group.

10. Do not try to assign all the bands in the spectrum. Fingerprint bands are unique to a particular system. Occasionally, intense bands will be fingerprint-type absorptions; these bands, generally, will be ignored in the interpretation. Fingerprint bands do, however, play an important role when infrared data are employed for identification purposes.

11. The correlation chart (back endpaper) can act as a helpful quick aid for checking potential assignments. It is not a substitute for understanding the theory and operation of group frequency logic. *The use of the correlation chart without a good knowledge of group frequencies is the shortest path to disaster!*[1]

12. Try to utilize the so-called *macro group frequency* approach. That is, if the functionality or molecular structural group requires the presence of more than a single group frequency vibrational mode, make sure that all modes are correctly represented. The *macro frequency train* represents a very powerful approach to the interpretation of relatively complex spectral data. Practice using *macro group frequencies* will pay big dividends in the laboratory. This last suggestion is perhaps the most important strategy to master in learning to interpret infrared spectra.

A SURVEY OF GROUP FREQUENCIES IDENTIFIED IN ORGANIC MOLECULES

The useful group frequencies are listed in the following sets of tables.

Note

> A detailed description of the associated fundamental vibrational modes, diagrams of the actual displacements of the atoms, along with associated spectra, may be found at ✹ Chapter 6, IR section, Part II. It is highly advisable to study this material.

In the following tables the vibration motion of the localized sections of the molecules assigned to a particular group's frequencies is often described using the following terms:

1. *Symmetric stretch* or *symmetric bend* (deformation): Here the local group retains its symmetry during displacement. The symmetric bend of the methylene group, CH_2, is often termed the *scissoring* bend, while the symmetric bend of a methyl group, CH_3, is termed an *umbrella* mode—both

---

[1]Bellamy, L. J. *The Infrared Spectra of Complex Molecules*, 3rd ed.; Chapman & Hall: London, 1975, p. 3.

descriptions imply the type of displacements that are taking place in the vibration.

2. *Antisymmetric stretch* or *bend* (deformation): Here the vibrating system loses its symmetry during the vibration. The displacements involve a reflective (mirror image) displacement during the opposite phase of the simple harmonic vibration and the motion is termed antisymmetric rather than asymmetric. Antisymmetric bends (deformations) are often classified as *twisting, rocking,* and *wagging* vibrations.

3. Some vibrations involving planar sections of the molecules are referred to as *in-plane* or *out-of-plane*. They can be either symmetric or antisymmetric in nature and if they involve the bending of all the displaced bonds of a set of atoms moving together in the same direction they will be termed *all-in-phase*.

4. *Degenerate* vibrations are defined as the case where two or more molecular vibrations are required to occur at the same frequency (see 🌀 Chapter 6, IR sections, Part I B and Part I D3 for more details).

5. *Overtones* (integral multiples of the fundamental mode frequency) and *sum tones* (the sum of two different fundamental modes) are forbidden bands that are almost always very weak. Occasionally these bands are good group frequencies.

Group frequencies of the hydrocarbons

Alkanes   Alkynes
Alkenes   Arenes

ALKANES

The C—H vibrational modes of the alkanes (or mixed compounds containing alkyl groups) that are characteristic and reliable group frequencies are summarized in Table 6.1 (also see 🌀 Chapter 6, IR section, Part II A).

These modes give rise to characteristic bands found in the infrared spectrum of alkanes, such as in the spectrum of *n*-hexane shown in Figure 6.2.

TABLE 6.1   *Alkane Vibrational Normal Modes*

C—H VIBRATIONAL MODES	$\bar{\nu} \pm 10$ (cm^{-1})
Methyl groups	
Antisymmetric (degenerate) stretch	2960
Symmetric stretch	2870
Antisymmetric (degenerate) deformation	1460
Symmetric (umbrella) deformation	1375
Methylene groups	
Antisymmetric stretch	2925
Symmetric stretch	2850
Symmetric deformation (scissors)	1450
Rocking mode (all-in-phase)	720

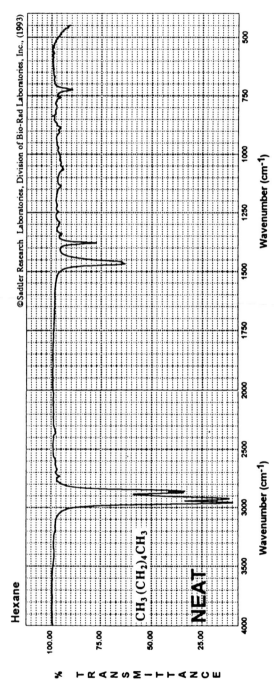

Figure 6.2 IR spectrum: *n*-hexane.

TABLE 6.2   *Substitution Classification of C=C Stretching Frequencies*

C=C NORMAL MODES	$\bar{\nu}$ (cm^{-1})
Trans-, tri-, tetrasubstituted	1680–1665
Cis-, vinylidene- (terminal 1,1-disubstituted), vinyl-substituted	1660–1620

ALKENES C=C STRETCHING

It is possible to classify open-chain unsaturated systems into two groups, those with C=C stretching modes falling above 1660 cm^{-1} and those with modes falling below 1660 cm^{-1} as shown in Table 6.2 (see also 🌐 Chapter 6, IR section, Part II B1).

ALKENES C—H

Several fundamental modes associated with the alkene C—H groups are group frequencies and are summarized in Table 6.3.

ALKYNES

The group frequencies of the alkynes are summarized in Table 6.4 (see also 🌐 Chapter 6, IR section, Part II C).

TABLE 6.3   *Alkene Vibrational Normal Modes*

C—H VIBRATIONAL MODES	$\bar{\nu} \pm 10$ (cm^{-1})
Stretching modes	
Antisymmetric stretch (=CH$_2$)	3080
Symmetric stretch (=CH$_2$)	3020
Uncoupled stretch (=CH)	3030
Out-of-plane bending modes	
Vinyl group	
Trans hydrogen atoms (in-phase)	990
Terminal hydrogen atoms (wag)	910
Vinylidene group (=CH$_2$)	
Terminal (wag)	890
Trans alkene	
Trans hydrogen atoms (in-phase)	965
Cis alkene	
Cis hydrogen atoms (in-phase)	~700
Trisubstituted alkene	
Uncoupled hydrogen atom	820
Tetrasubstituted alkene: no vibrational modes seen in IR	

TABLE 6.4 *Alkyne Vibrational Normal Modes*

C≡C, C—H VIBRATIONAL MODES	$\bar{\nu} \pm 10$ (cm^{-1})
Triple-bond stretch (monosubstituted)	2120
Triple-bond stretch (disubstituted)	2225
R—C≡C—H bond stretch (monosubstituted)	3300

ARENES

The group frequencies of the *phenyl* group can be classified as carbon–hydrogen vibrations consisting of stretching and out-of-plane bending modes, plus carbon–carbon ring stretching and out-of-plane bending modes. The in-plane bending modes in both cases are not effective group frequencies.

The wavenumber values for the all-in-phase, C—H bending vibrations are presented in Table 6.5.

The generalized group frequencies of the arenes are summarized in Table 6.6 (see also 🌐 Chapter 6, IR section, Part II D).

TABLE 6.5 *Arene Out-of-Ring-Plane C—H Deformation Modes*

ARENE FUNDAMENTALS (C—H BEND) (NUMBER OF C—H GROUPS DIRECTLY ADJACENT)	$\bar{\nu}$ RANGE (cm^{-1})
5	770–730
4	770–735
3	810–750
2	860–800
1	900–845

Group frequencies of carbonyl groups: C=O

The carbonyl group is perhaps the single most important functional group in organic chemistry. It is certainly the most commonly occurring functionality. Infrared spectroscopy can play a powerful role in the characterization of the carbonyl because this group possesses all of the properties that give rise to an excellent group frequency. (Table 6.7; for an in depth-discussion see 🌐 Chapter 6, IR section, Part III A.)

The major factors perturbing carbonyl frequencies can be summarized as follows:

FACTORS THAT RAISE THE C=O FREQUENCY

1. Substitution with electronegative atoms

2. Decrease in C—CO—C internal bond angle

TABLE 6.6 *Arene Group Frequencies*

ARENE FUNDAMENTALS	$\tilde{\nu}$ RANGE (cm^{-1})
C—H stretch	3100–3000
C=C ring stretch (ν_{8a})	1600 ± 10
C=C ring stretch (ν_{8b})	1580 ± 10
C=C ring stretch (ν_{19a})	1500 ± 10
C=C ring stretch (ν_{19b})	1450 ± 10
C—H out-of-plane bend (1H)	900–860
C—H out-of-plane bend (2H)	860–800
C—H out-of-plane bend (3H)	810–750
C—H out-of-plane bend (4H)	770–735
C—H out-of-plane bend (5H)	770–730
C—C ring out-of-plane bend (1; 1,3; 1,3,5-substituted)	690 ± 10
C—H out-of-plane bend sum tones	2000–1650

FACTORS THAT LOWER THE C=O FREQUENCY

1. Conjugation

2. Hydrogen bonding

Several of these factors may be operating simultaneously, so careful judgment as to the contribution of each individual effect must be exercised in predicting carbonyl frequencies. This judgment develops rapidly with practice at interpretation.

TABLE 6.7 *Carbonyl Group Vibrational Frequencies*

COMPOUND	$\tilde{\nu}$ (cm^{-1})
Ketones, aliphatic, open-chain (R$_2$CO)	1725–1700
Ketones, conjugated	1700–1675
Ketones, cyclic	*a*
Acyl halides	>1800
Esters, aliphatic	1755–1735
Esters, conjugated	1735–1720
Esters (conjugated to oxygen)	1780–1760
Lactones	*a*
Anhydrides: aliphatic, open-chain	1840–1810 and 1770–1740
Carboxylic acids, aliphatic	1725–1710
Amides	(see Table 6.19–6.21)
Lactams	*a*
Aldehydes	1735–1720

^{a}see Chapter 6, IR section, Part III A.

TABLE 6.8 *Vibrational Normal Modes of the Hydroxyl Group*

$\bar{\nu}$ (cm^{-1})	INTENSITY	MODE DESCRIPTION
3500–3200	Very strong	O—H stretch (only strong when hydrogen bonded)
1500–1300	Medium strong	O—H in-plane bend (overlap CH_2, CH_3 bend)
1260–1000	Strong	C—C—O antisymmetric stretch
650	Medium	O—H out-of-plane bend

Group frequencies of the heteroatom functional groups

(Alkanes)	Carboxylic acids	Halogens
Alcohols	Nitriles	Esters
Acyl halides	Thiols	Ethers
Amines, primary	Ketones	Amides, secondary
Isocyanates	Anhydrides	Phenyl
Aldehydes	Amides, primary	

HEXANE
Refer to Table 6.1 (see also 🌐 Chapter 6, IR section, Part II A, and Fig. 6.2).

ALCOHOLS
A very intense band appears at ~3350 cm^{-1}, which is assigned to the stretching mode of the O—H group (Table 6.8; also see 🌐 Chapter 6, IR section, Fig. W6.24).

Of particular importance is a strong band in the spectrum of aliphatic alcohols usually located near 1060 cm^{-1}. This absorption has been identified as the C—O stretching mode. The vibrational displacements of this fundamental are similar to the antisymmetric stretch of water (see 🌐 Chapter 6 for a detailed discussion of the vibrational modes of the water molecule). Since the vibration involves significant displacement of the adjacent C—C oscillator, the vibration will be substitution sensitive. These latter shifts can be of value in determining the nature of the alcohol (primary, secondary, or tertiary, see Table 6.9).

TABLE 6.9 *Substitution Effects on C—O Stretch of Aliphatic Alcohols*

TYPE OF ALCOHOL	$\bar{\nu}_{C-O}$ (cm^{-1})
RCH_2—OH (primary)	1075–1000
R_2CH—OH (secondary)	1150–1075
R_3C—OH (tertiary)	1200–1100
C_6H_5—OH (phenol)	1260–1180

TABLE 6.10   *Vibrational Normal Modes of the Aliphatic Aldehyde Group*

$\bar{\nu}$ (cm^{-1})	INTENSITY	MODE DESCRIPTION
2750–2720	Weak to medium	C(O)—H stretch (see also 🌐, Fermi resonance discussion)
1735–1720	Very strong	C=O stretch
1420–1405	Medium	CH$_2$ symmetric bend, —CH$_2$— α to —CHO
1405–1385	Medium	C—H in-plane bend

ALDEHYDES

The aldehydes functional groups gives rise to several good group frequencies Table 6.10; also see 🌐 Chapter 6, IR section, Fig. W6.25).

KETONES

The only group frequency mode associated directly with aliphatic ketones is the stretching frequency ($\tilde{\nu}_{C=O}$ ~1720 cm^{-1}), which occurs within the expected region as discussed above. There are, however, several other related bands (Table 6.11; also see 🌐 Chapter 6, IR section, Fig. W6.26).

TABLE 6.11   *Normal Vibrational Modes of Aliphatic Ketones*

$\bar{\nu}$ (cm^{-1})	INTENSITY	MODE DESCRIPTION
3430–3410	Very weak	Overtone of carbonyl stretch
1725–1700	Very strong	C=O
1430–1415	Medium	—CH$_2$— symmetric bend, —CH$_2$— α to ketone C=O

ESTERS

The very strong band found at ~1745 cm^{-1} is typical of the carbonyl frequency of an aliphatic ester, particularly aliphatic acetate esters (Table 6.12; also see 🌐 Chapter 6, IR section, Fig. W6.27).

ACYL HALIDES

The carbonyl stretching mode dominates the spectrum in aliphatic acyl halides. In acyl chlorides it is an extremely intense band occurring near 1800 cm^{-1} (Table 6.13; also see 🌐 Chapter 6, IR section, Fig. W6.28).

CARBOXYLIC ACIDS

Acids, observed in the solid or pure liquid states, often possess a very intense band with a width at one-half peak height of about 1000 cm^{-1}, which covers the region 3500–2200 cm^{-1}. This absorption is characteristic of very

TABLE 6.12 *Vibrational Normal Modes of the Aliphatic Ester Group*

$\tilde{\nu}$ (cm^{-1})	INTENSITY	MODE DESCRIPTION
1755–1735	Very strong	C=O stretch
1370–1360	Medium	CH$_3$ symmetric bend α to ester C=O
1260–1230	Very strong	C—CO—O antisymmetric stretch —acetates
1220–1160	Very strong	C—CO—O antisymmetric stretch —higher esters
1060–1030	Very strong	O—CH$_2$—C antisymmetric stretch —1° acetates
1100–980	Very strong	O—CH$_2$—C antisymmetric stretch —higher esters (may overlap with upper band)

TABLE 6.13 *Vibrational Normal Modes of the Acyl Halide Group*

$\tilde{\nu}$ (cm^{-1})	INTENSITY	MODE DESCRIPTION
1810–1800	Very strong	C=O stretch, acyl chlorides
1415–1405	Strong	—CH$_2$— symmetric bend, α to —COCl carbonyl

strongly hydrogen-bonded carboxylic acid groups (Table 6.14; also see Chapter 6, IR section, Fig. W6.29).

TABLE 6.14 *Vibrational Normal Modes of the Carboxylic Acid Group*

$\tilde{\nu}$ (cm^{-1})	INTENSITY	MODE DESCRIPTION
3500–2500	Very very strong	O—H stretch intensified by hydrogen bonding
2800–2200	Very weak	Overtone and sum tones
1725–1710	Very strong	C=O antisymmetric hydrogen-bonded dimer stretch
1450–1400	Strong	CH$_2$—CO—O antisymmetric stretch mixed with O—H bend
1300–1200	Strong	CH$_2$—CO—O antisymmetric stretch mixed with O—H bend
950–920	Medium	Out-of-plane O—H bend, acid dimer

ANHYDRIDES

The coupling of the anhydride carbonyls through the ether oxygen splits the carbonyls (in the aliphatic case $\tilde{\nu}_{C=O} = {\sim}1830, 1760$ cm^{-1}) by about 70 cm^{-1} (Table 6.15; also see Chapter 6, IR section, Fig. W6.30).

TABLE 6.15 *Vibrational Normal Modes of the Anhydride Group*

$\tilde{\nu}$ (cm^{-1})	INTENSITY	MODE DESCRIPTION
1840–1810	Very strong	C=O in-phase stretch
1770–1740	Very strong	C=O out-of-phase stretch
1420–1410	Strong	—CH$_2$— symmetric bend α to C=O
1100–1000	Very strong	C—O stretch, mixed modes

ETHERS

The large intensity associated with antisymmetric C—O—C stretching modes relative to the other bands occurring in this part of the fingerprint region, particularly in aliphatic compounds, makes it possible, in most cases, to assign with confidence the observed strong band (Table 6.16; also see 🌐 Chapter 6, IR section, Fig. W6.31).

TABLE 6.16 *Vibrational Normal Modes of the Ether Group*

$\tilde{\nu}$ (cm^{-1})	INTENSITY	MODE DESCRIPTION
1150–1050	Strong	C—O—C antisymmetric stretch, mixed mode

PRIMARY AMINES

The spectra of these bases usually possess two bands ($\tilde{\nu}_{N—H}$ = ~3380, ~3300 cm^{-1}) of medium-to-weak intensity. These bands are assigned to the antisymmetric and symmetric N—H stretching modes, respectively, of the primary amino group (Table 6.17; also see 🌐 Chapter 6, IR section, Fig. W6.32).

TABLE 6.17 *Vibrational Normal Modes of the Primary Amine Group*

$\tilde{\nu}$ (cm^{-1})	INTENSITY	MODE DESCRIPTION
3400–3200	Weak to medium	NH$_2$ stretch doublet, (antisymmetric and symmetric modes)
1630–1600	Medium	NH$_2$ symmetric bend
820–780	Medium	NH$_2$ wag

NITRILES

The very strong triple bond present in the nitrile group (as in the case of the alkynes) contributes to an unusually high stretching frequency, and the polar character of the group gives rise to very intense bands (Table 6.18; also see 🌐 Chapter 6, IR section, Fig. W6.33)

PRIMARY AMIDES

The highly polar amide group leads to very strong hydrogen bonding, which in turn leads to greatly intensified N—H antisymmetric and symmetric

TABLE 6.18   *Vibrational Normal Modes of the Nitrile Group*

$\tilde{\nu}$ (cm^{-1})	INTENSITY	MODE DESCRIPTION
2260–2240	Strong	C≡N stretch, aliphatic
2240–2210	Strong	C≡N stretch, conjugated

stretching modes ($\tilde{\nu}_{\text{N—H}}$ = ~3375, ~3200 cm^{-1}; see Table 6.19; also see ⬤ Chapter 6, IR section, Fig. W6.34).

TABLE 6.19   *Vibrational Normal Modes of the Primary Amide Group*

$\tilde{\nu}$ (cm^{-1})	INTENSITY	MODE DESCRIPTION
3400–3150	Very strong	—NH$_2$ antisymmetric and symmetric stretching modes, hydrogen bonded
1680–1650	Very strong	C=O stretch, hydrogen bonded
1660–1620	Strong	—NH$_2$ symmetric bend (overlap with C=O stretch)
1430–1410	Strong	—CH$_2$— symmetric bend α to amide carbonyl
750–650	Medium	—NH$_2$ wag

SECONDARY AMIDES

The single N—H group present in secondary amides gives rise to a very strong band near about 3300 cm^{-1}, which is indicative of strong hydrogen bonding. The drop in frequency from that of the primary —NH$_2$ scissoring mode near 1600 cm^{-1} allows for confident assignment of substitution on secondary amide groups (Table 6.20; also see ⬤ Chapter 6, IR section, Fig. W6.35).

Studies of amide carbonyl frequencies in dilute nonpolar solution indicate that hydrogen-bonding effects are largely responsible for the low frequencies observed with primary and secondary amides, but play no role in tertiary amides (see Table 6.21).

TABLE 6.20   *Vibrational Normal Modes of the Secondary Amide Group*

$\tilde{\nu}$ (cm^{-1})	INTENSITY	MODE DESCRIPTION
3350–3250	Strong	—NH stretch, intensified by hydrogen bonding
3125–3075	Medium	Overtone N—H bend (see also ⬤ Fermi resonance discussions)
1670–1645	Very strong	C=O stretch, hydrogen bonded
1580–1550	Strong	N—H in-plane bend (see also ⬤ coupling discussions)
1415–1405	Strong	—CH$_2$— symmetric bend α to amide C=O
1325–1275	Medium	C—N stretch mixed with N—H in-plane bend
725–680	Medium	N—H out-of-plane bend

TABLE 6.21 *Vibrational Normal Modes of the Amide Carbonyl: Solution and Solid-Phase Data*

AMIDE	DILUTE SOLUTION (cm^{-1})	SOLID (cm^{-1})
R—CO—NH$_2$ (primary)	~1730	~1690–1650
R—CO—NHR (secondary)	~1700	~1670–1630
R—CO—NR$_2$ (tertiary)	~1650	~1650

ISOCYANATES

The range of stretching frequencies observed for alkyl-substituted isocyanates is very narrow, $\tilde{\nu} = 2280$–2260 cm^{-1}, which implies little coupling to the rest of the system (Table 6.22; also see ⬤ Chapter 6, IR section, Fig. W6.36).

TABLE 6.22 *Vibrational Normal Mode of the Isocyanate Group*

$\tilde{\nu}$ (cm^{-1})	INTENSITY	MODE DESCRIPTION
2280–2260	Very strong	—N=C=O antisymmetric stretch

THIOLS

Although weak absorption is associated with the S—H stretching fundamental the band is generally found in a very open region of the infrared spectrum (Table 6.23; also see ⬤ Chapter 6, IR section, Fig. W6.37).

TABLE 6.23 *Vibrational Normal Mode of the Thiol Group*

$\tilde{\nu}$ (cm^{-1})	INTENSITY	MODE DESCRIPTION
2580–2560	Weak	S—H stretch

ALKYL HALIDES

The massive halogen atom is connected to the alkyl section by a fairly weak but highly polarized bond, which dictates that the C—X stretching frequency appears as an intense band at low frequencies (Table 6.24; also see ⬤ Chapter 6, IR section, Fig. W6.38).

TABLE 6.24 *Vibrational Normal Mode of the Alkyl Chloro Group*

$\tilde{\nu}$ (cm^{-1})	INTENSITY	MODE DESCRIPTION
750–650	Strong	C—Cl stretch (see also ⬤)

ARYL HALIDES (CHLOROBENZENE)

The final system to be considered in this section is the aryl halide, chlorobenzene. Based on the above assignments the group frequencies of the complete hydrocarbon portion and the heteroatom functional group can be assigned as in Table 6.25 (also see ⬤ Chapter 6, IR section, Fig. W6.39).

TABLE 6.25 *Vibrational Normal Modes of the Aryl Chloro Group*

$\tilde{\nu}$ (cm^{-1})	INTENSITY	MODE DESCRIPTION
3080	Medium	C—H stretch, bonded to ring carbon
1585	Strong	ν_{8a} ring stretching
1575	Weak	ν_{8b} ring stretching
1475	Strong	ν_{19a} ring stretching
1450	Strong	ν_{19b} ring stretching
747	Strong	C—H all-in-phase, out-of-plane bend
700	Strong	C—Cl stretch
688	Strong	Ring deformation
1945, 1865, 1788, 1733	All weak	Sum tones, out-of-plane C—H bends, pattern matches monosubstitution of ring

INFRARED SPECTROSCOPY INSTRUMENTATION AND SAMPLE HANDLING

Instrumentation

The workhorse infrared instrument used for routine characterization of materials in the undergraduate organic laboratory is still the optical-null double-beam grating spectrometer (Fig. 6.3). For a discussion of double-beam spectrometers, see the UV-vis instrumentation discussion (p. 231). Although many undergraduate instructional laboratories still utilize this type of instrumentation, the winds of change are blowing. Lower-cost FT-IR spectrometers, which depend on high-speed computer manipulation of the spectral data, are becoming the infrared instructional instrumentation of choice. The spectra utilized in the infrared analyses (pp. 166–170) were generated on a prototype of this kind of infrared instrumentation, the Perkin–Elmer model 1600.

This instrument can acquire 16 scans and carry out the required calculations in 42 s. While the spectrum is being printed out (~40 s) the data on a second sample can be acquired. The 42-s acquisition data are significantly superior to those currently recorded by dispersive instruments that take from 5 to 8 min to scan a sample from 4000 to 600 cm^{-1}.

A short description of FT-IR spectrometers is included in the discussion of instrumentation on the website (⬤ Chapter 6, IR section, Part IV).

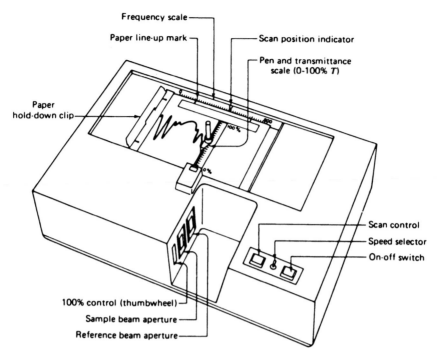

Figure 6.3 The Perkin–Elmer model 710B IR spectrometer. *From Zubric, James W.* The Organic Chem Lab Survival Manual, *4th ed.; Wiley: New York, 1997. (Reprinted by permission of John Wiley & Sons, Inc., New York.)*

Sample handling in the infrared

The standard techniques of sample preparation employed to obtain infrared spectra of microscale laboratory products are the use of capillary films with liquids on NaCl (or AgCl) plates and the use of KBr disks and melts in solids. For a spectrum to be obtained in the infrared region, the sample must be mounted in a cell that is transparent to the radiation. Glass and quartz absorb in this spectral region, so cells constructed of these materials cannot be used. Alkali metal halides have large spectral regions of transmission in the infrared, as do silver halides. Sodium chloride is the most commonly used material in cell windows in infrared sampling.

LIQUID SAMPLES

For materials boiling above 100 °C, the procedure is very simple. Using a syringe or Pasteur pipet, place 3–5 µL of sample on a polished plate of sodium chloride or silver chloride. Cover it with a second plate of the same material and clamp it in a holder that can be mounted vertically in the instrument. Be sure that the plates are clean when you start and when you are through! Obviously, the sodium chloride plates cannot be cleaned with water. Silver chloride is very soft and scratches easily; it also must be kept in the dark when not in use because it darkens quickly in direct light. Spectra obtained in this fashion are referred to as *capillary film spectra* (Fig. 6.4).

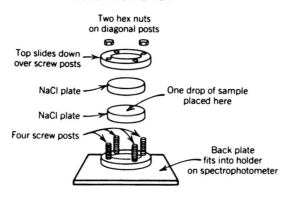

DO NOT OVERTIGHTEN

Figure 6.4 IR salt plates and holder. *From Zubric, James W.* The Organic Chem Lab Survival Manual, *4th ed.;* Wiley: New York, 1997. *(Reprinted by permission of John Wiley & Sons, Inc., New York.)*

SOLUTION SPECTRA AND THE SPECTRA OF MATERIALS BOILING BELOW 100 °C

These samples generally require a sealed cell constructed of either sodium chloride or potassium bromide windows. Such cells are expensive and need careful handling and maintenance. They are assembled as shown in Figure 6.5.

SOLID SAMPLES

Solid powders could be mounted on horizontal sodium chloride plates, and the beam diverted through the sample by mirrors. This procedure would make sample preparation very easy for solids. Unfortunately, powders tend to scatter the entering radiation very efficiently by reflection, refraction, and molecular scattering. Some of these effects become rapidly magnified at higher frequencies, since they vary with the fourth power of the frequency. Thus, in solid-sample irradiation a lot of energy is scattered away from the sample beam. This results in poor absorption spectra, as the instrument is forced to operate

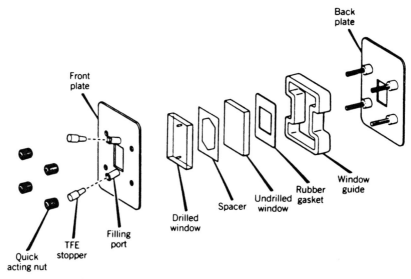

Figure 6.5 Sealed demountable cell or demountable cell with ports. *(Courtesy of the Perkin–Elmer Corp., Norwalk, CT.)*

at very low energies. The detector cannot differentiate between a drop in energy from absorption or one derived from scattering.

For materials melting below 80 °C the simplest technique is to mount the sample between two salt plates and **gently** warm with a heat lamp until melting occurs. With the fast acquisition times of interferometers, the melting point range is now as high as 100 °C and the spectrum can be obtained so rapidly that the sample does not have time to cool and crystallize. (Heated cells are used in research laboratories, but they are rather expensive and difficult to maintain.) For substances melting above 100 °C, the sampling routine most often employed to avoid scattering problems is the potassium bromide (KBr) disk. Potassium bromide is transparent to infrared radiation in the region of interest. Most important, however, the KBr makes a much better match of the refractive indexes between the sample and its matrix than does air. Thus, reflection and refraction effects at the crystal faces of the sample are greatly suppressed.

In the KBr method the sample (2–3 mg) is finely ground in a mortar, the finer the better for lower reflection or refraction losses. Then, 150 mg of previously ground and dried KBr is added to the mortar and quickly mixed by stirring, not grinding, it with the sample. (Potassium bromide is very hygroscopic and will rapidly pick up water while being ground in an open mortar.) When mixing is complete the mixture is transferred to a die and pressed into a solid disk. Potassium bromide will flow under high pressure and seal the solid sample in a glasslike matrix. Several styles of dies are commercially available. For routine use a die consisting of two stainless steel bolts and a barrel is the simplest to operate (see Fig. 6.6). The ends of the bolts are polished flat to form the die faces. The first bolt is seated to within a turn or two of the head. Then the sample mixture is added (avoid breathing over the die while adding the sample). The second bolt is firmly seated in the barrel, and then the clamped assembly is tightened by a torque wrench to 240 in./lb. After standing for 1.5 min, the two bolts are removed, leaving the KBr disk mounted in the center of the barrel, which can then be mounted in the instrument. After the spectrum of the sample is run, the disk can be retrieved and the sample recovered if necessary (Fig. 6.6). *Always clean the die immediately after use. KBr is highly corrosive to steel.*

When infrared spectra are obtained, it is important to establish that the wavenumber values have been accurately recorded. Successful interpretation of the data often depends on very small shifts in these values. Calibration of the frequency scale is usually accomplished by obtaining the spectrum of a reference compound, such as polystyrene film. To save time, record absorption peaks only in the region of particular interest (this applies only to dispersive instrument derived data).

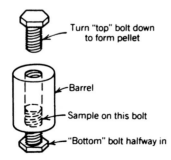

Turn "top" bolt down to form pellet

Barrel

Sample on this bolt

"Bottom" bolt halfway in

Figure 6.6 The KBr pellet minipress. *From Zubric, James W.* The Organic Chem Lab Survival Manual, *4th ed.: Wiley: New York, 1997. (Reprinted by permission of John Wiley & Sons, Inc., New York.)*

Note _____

> It should be noted that the infrared spectra referred to in the examples are Fourier-transform-derived and have been plotted on a slightly different scale than the other spectra presented in the text and on the website. These spectra utilize a 12.5-cm^{-1}/mm format below 2000 cm^{-1} and undergo a 2:1 compression above 2000 cm^{-1}(25-cm^{-1}/mm).

Infrared Analysis of the Reduction of Ketones Using a Metal Hydride Reagent: Cyclohexanone to Cyclohexanol

This example illustrates the power of infrared spectroscopy to follow the course of an organic reaction: the reduction of a ketone (cyclohexanone) carbonyl to the corresponding alcohol (cyclohexanol) by use of sodium borohydride.[2]

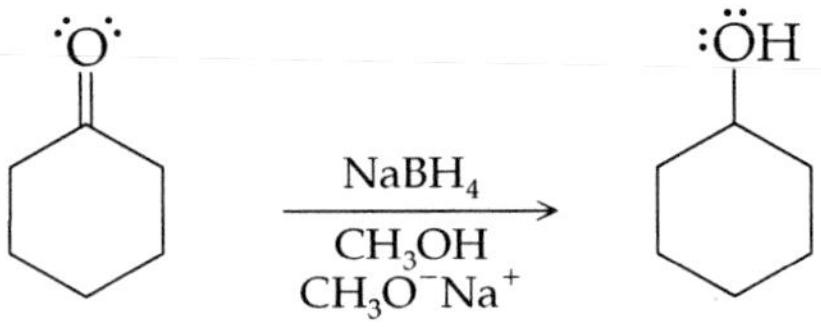

Cyclohexanone $\xrightarrow[\substack{CH_3OH \\ CH_3O^-Na^+}]{NaBH_4}$ Cyclohexanol

A SPECTRAL COMPARISON OF REACTANT AND PRODUCT

The key absorption bands to examine in the spectrum of cyclohexanone occur at 3420, 3000 to 2850, 1715, and 1425 cm^{-1} (Fig. 6.7). The lack of significant absorption between 3100 to 3000 and 1400 to 1350 cm^{-1} also should be noted. The sharp weak band at 3420 cm^{-1} is not a fundamental vibration (not O—H or N—H stretching), but arises from the first overtone of the very intense carbonyl stretching mode found at 1715 cm^{-1}. Note that the overtone does not fall exactly at double the frequency of the fundamental, but usually occurs slightly below that value because of anharmonic effects.

The absence of absorption in the region near 3100 to 3000 cm^{-1} and the presence of a series of very strong absorption bands at 3000 to 2850 cm^{-1} in-

---

[2]Mayo, D. W.; Pike, R. M.; Trumper, P. K. *Microscale Organic Laboratory*, 4th ed., Wiley; New York, 2000, p. 136.

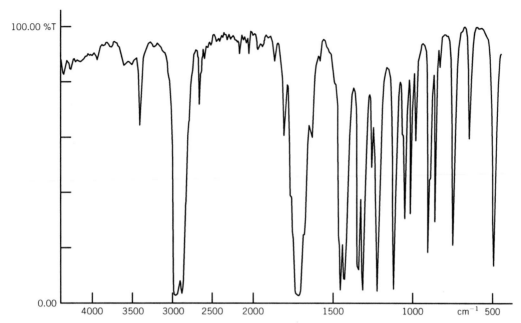

Figure 6.7 IR spectrum: cyclohexanone.

dicate that the only C—H stretching modes present are part of sp^3 type systems. Thus, the spectrum is typical of an aliphatic ketone. The occurrence of a band at 1425 cm^{-1} suggests the presence of at least one methylene group adjacent to the carbonyl group, whereas the 1450 cm^{-1} band requires other methylene groups more remote from the C=O group. The lack of absorption in the 1400 to 1375 cm^{-1} region indicates the absence of any methyl groups (a good indication of a simple aliphatic ring system) and that the absorption at 1450 cm^{-1} must arise entirely from methylene scissoring modes. The value of 1715 cm^{-1} for the C=O stretch supports the presence of a six-membered ring.

Now examine the spectrum of the cyclohexanol reaction product (a typical example is given in Fig. 6.8).

The spectrum is rather different from that of the starting material. The major differences are a new very strong broad band occurring between 3500 and 3100 cm^{-1} and a great drop in intensity of the band found at 1715 cm^{-1}. These changes indicate the reductive formation of an alcohol group from the carbonyl system. The band centered near 3300 cm^{-1} results from the single highly polarized O—H stretching mode. The drop in intensity of the 1715 cm^{-1} band indicates the loss of the carbonyl function.

The exact amount of cyclohexanone remaining could very easily be determined by carrying out a Beer's law type of analysis. The transmitted beam undergoes a certain amount of instrument distortion, but generally follows the Beer–Lambert absorption law:

$$I/I_0 = e^{-acl}$$

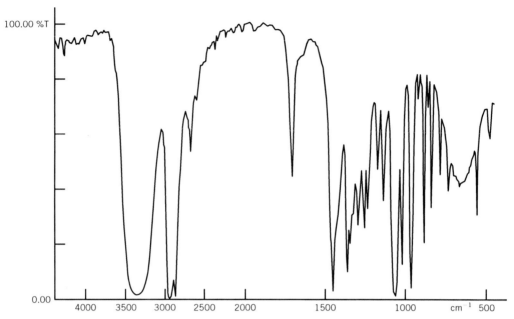

Figure 6.8 IR spectrum: cyclohexanol (reaction product).

where I = intensity of the transmitted beam

I_0 = intensity of incident beam

a = apparent absorption coefficient

c = sample concentration, mol/L

l = path length (sample thickness)

Other bands of interest in the spectrum of cyclohexanol occur at 1069 and 1031 cm^{-1}. These can be assigned respectively to the equatorial and axial C—O stretching of the rotational conformers of this alicyclic secondary alcohol. A broad band (width ~300 cm^{-1}) can be found near 670 cm^{-1} and arises from an O—H bending, out-of-plane mode of the associated alcohol. This band is generally identified only in neat samples where extensive H-bonding occurs. Also note that the band at 1425 cm^{-1} has vanished as there are no methylene groups α to carbonyl systems in the product.

An Additional Example of Infrared Analysis: Reduction of t-Butylcyclohexanone

This example further illustrates the value of infrared spectroscopy in following the reduction of a ketone (*t*-butylcyclohexanone) to the corresponding isomeric alcohols (*cis*-, *trans-t*-butylcyclohexanols) using a metal hydride.[3]

[3]Mayo, D. W.; Pike, R. M.; Trumper, P. K. *Microscale Organic Laboratory*, 4th ed., Wiley; New York, 2000, p. 140.

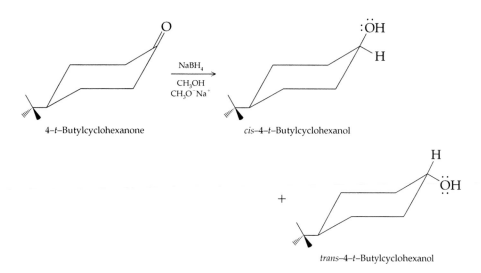

Refer to the discussion in the preceding example for an interpretation of the absorption bands found at 3435, 3000 to 2850, 1715, and 1425 cm^{-1} in the starting material (Fig. 6.9), and at 3250, 3000 to 2850, 1715 (variable relative intensity—may be quite weak—why?), 1069, and 1031 cm^{-1} in the crude alcohol (Fig. 6.10). The *t*-butyl-ketone has, in addition, bands at 1396 (weak) and 1369 (strong) cm^{-1}, and the alcohol has bands at 1399 (weak) and 1375 (strong) cm^{-1}. These two pairs of bands establish the presence of the tertiary butyl group in these compounds.

Note that (1) a weak band (3495 cm^{-1}) is present on the high wavenumber side of the 3250 cm^{-1} O—H stretching mode and (2) even in neat samples

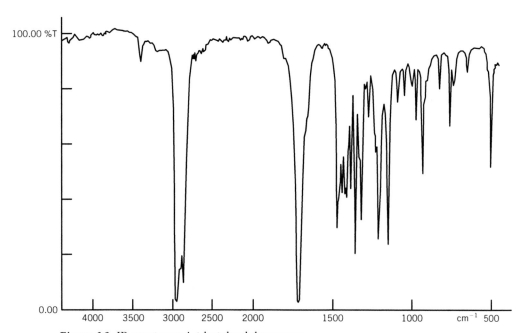

Figure 6.9 IR spectrum: 4-*t*-butylcyclohexanone.

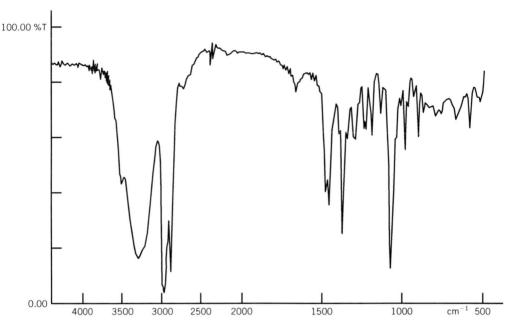

Figure 6.10 IR spectrum: 4-*t*-butylcyclohexanol.

of the tertiary butyl derivative, the 670 cm^{-1} band, clearly evident in cyclo-hexanol, is difficult to observe.

See Chapter 7 for a detailed discussion of the NMR analysis of these diastereoisomeric products. This illustrates the value of the intergration of NMR and IR spectroscopic techniques. The synergistic effect of combining these techniques can provide a powerful means of following the course of this particular metal hydride reduction and chemical reactions in general.

QUESTIONS

Note

Some of the following questions assume that the student is familiar with the infrared material contained on the 🌐 reference or at least has the website available to refer to if needed.

6-1. The form of the C—H out-of-plane bending vibrations of the vinyl group are shown on the following page:
The first two vibrational modes give rise to excellent group frequencies, while the third fundamental does not lend itself to these correlations.

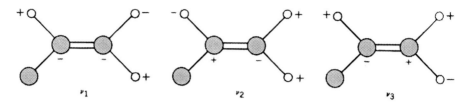

(a) Explain the factors that lead to the third vibrational mode being such a poor group frequency.

(b) Predict the location in the spectrum of the third fundamental vibration.

6-2. In the figure below, the mass of the terminal hydrogen atoms on the acetylene is hypothetically varied from zero to infinity. The response of the C—H symmetric stretching (3374 cm^{-1}) and triple-bond stretching (1974 cm^{-1}) modes to the change in mass is shown.

(a) Calculate the expected deuterium isotope shift for the C—H symmetric stretching mode. Is the hypothetical value close to the calculated value? Explain.

(b) Explain why the triple-bond stretching frequency is approximately 100 cm^{-1} higher for high-mass terminal isotopes ($>$100) than for the low-mass terminal isotopes ($<$3).

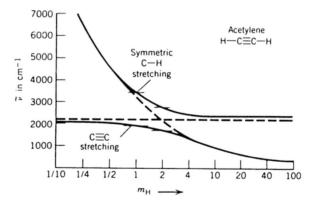

6-3. Acetylene has two C—H groups. It will have two C—H stretching frequencies, the in-phase and out-of-phase stretching modes. The in-phase (symmetric) stretch occurs at 3374 cm^{-1} and the out-of-phase stretch at 3333 cm^{-1}. Explain why the in-phase vibration is located at a higher frequency than the out-of-phase stretch.

6-4. The carbonyl stretching frequencies of a series of benzoyl derivatives are listed below:

x	$\tilde{\nu}_{C=O}$ (CCl$_4$)
2	1677
3	1686
4	1687
5	1686

Consider the $\tilde{\nu}_{C=O}$ of acetone at 1715 cm^{-1} as a reference frequency and identify the factors affecting $\tilde{\nu}_{C=O}$ in the series of compounds listed.

6-5. Explain how mass effects act to lower the carbonyl frequency, as well as how inductive and hyperconjugation effects act to raise the carbonyl frequency of aldehydes relative to ketones.

6-6. The carbonyl stretching frequency of aliphatic carboxylic acids in dilute solution is located near 1770 cm^{-1}. This frequency is much higher than the carbonyl frequency of these substances when measured neat ($\sim$1720 cm^{-1}). Also, it is considerably higher than the corresponding simple aliphatic ester value (1745 cm^{-1}). Explain.

6-7. In a number of cases, dipolar interactions control the frequency shifts found in carbonyl stretching vibrations. The table lists wavenumber shifts in going from neat to dilute nonpolar solutions. Explain the observed values.

Carbonyl Dipolar Interactions[a]

COMPOUND	$\Delta \tilde{\nu}$ (cm^{-1})
Acetyl chloride	15
Phosgene	13
Acetone	21
Acetaldehyde	23
N,N-Dimethylformamide	50

[a]Shift measured between dilute nonpolar solution and neat sample.

6-8. The antisymmetric —CH$_2$—CO—O— stretching vibration in carboxylic acids is heavily mixed with the in-plane bending mode of the O—H group. In alcohols these two vibrations seldom show evidence of mechanical coupling. Explain.

6-9. Conjugation of the functional group in alkyl isocyanates has little impact on the antisymmetric —N=C=O stretching vibration located near 2770 cm^{-1}. Explain.

6-10. In the infrared spectrum of 2-aminoanthraquinone (**I**) two carbonyl stretching frequencies are observed at 1673.5 and 1625 cm^{-1}:

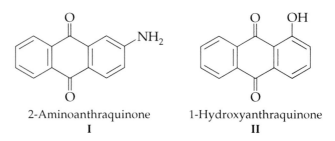

2-Aminoanthraquinone
I

1-Hydroxyanthraquinone
II

(a) Assign carbonyl bands in the infrared spectrum to the carbonyl groups in structure **I** and explain your reasoning.

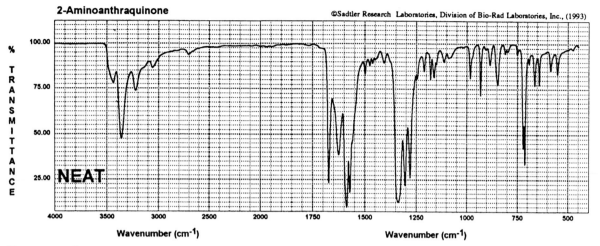

IR spectrum: 2-Aminoanthraquinone.

(b) The infrared spectrum of 1-hydroxyanthraquinone (**II**) also exhibits two carbonyl frequencies, which are located at 1675 and 1637 cm^{-1}. Assign the carbonyl groups to the related absorption bands. Explain your reasoning.

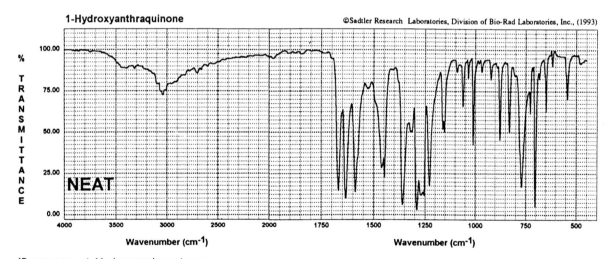

IR spectrum: 1-Hydroxyanthraquinone.

(c) The spectrum of 2-hydroxyanthraquinone exhibits a single carbonyl stretching frequency near 1673 cm^{-1}. Explain why a single carbonyl band would be expected in the system and why this vibration is located at 1673 cm^{-1}.

6-11. Suggest a possible structure for the hydrocarbon C_6H_{14}, which has the infrared spectrum shown on the next page: Is there more than one possible correct structure?

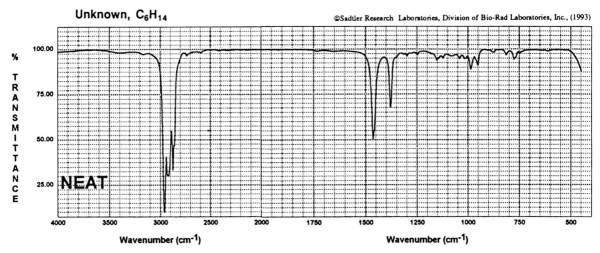

IR unknown spectrum: C_6H_{14}.

6-12. The hydroxylamine **I** can be oxidized by MnO_2 to the amide oxohae-manthidine (**II**). In dilute solution the carbonyl absorption band of **II** occurs at 1702 cm^{-1}. Explain this observation.

6-13. Identify the following alkenes (*a–e*). All samples were obtained from distillation cuts in the C_6 boiling range.

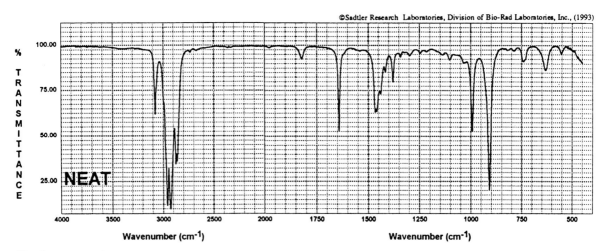

IR unknown spectrum *a*.

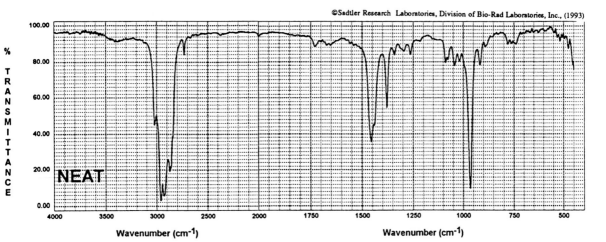

IR unknown spectrum *b*.

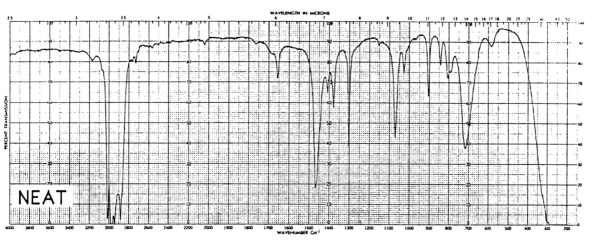

IR unknown spectrum *c*. *(Courtesy of Bowdoin College)*

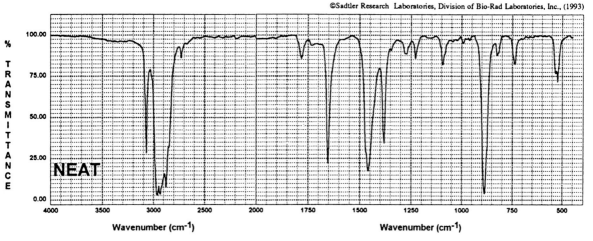

IR unknown spectrum *d*.

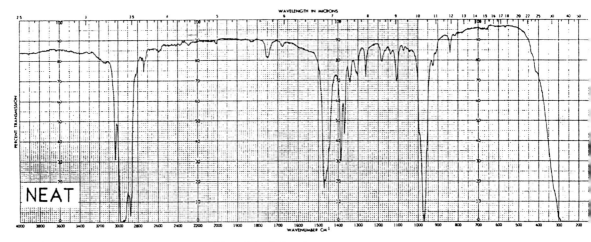

IR unknown spectrum *e.* *(Courtesy of Bowdoin College)*

6-14. The infrared spectra of the three xylene (dimethylbenzene) isomers, and an additional aromatic hydrocarbon, are given below. Assign the spectra to the isomers and suggest a potential structure for the remaining unknown substance.

©Sadtler Research Laboratories, Division of Bio-Rad Laboratories, Inc., (199?

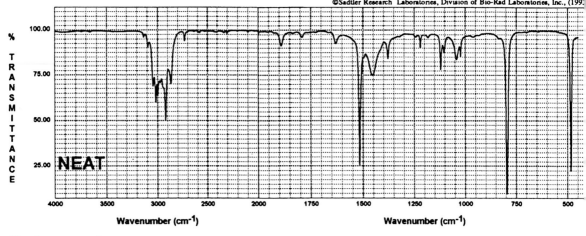

IR unknown spectrum *a.*

©Sadtler Research Laboratories, Division of Bio-Rad Laboratories, Inc., (199

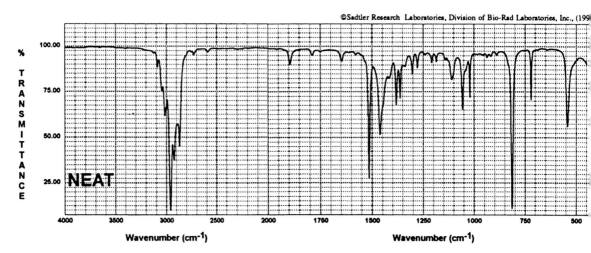

IR unknown spectrum *b.*

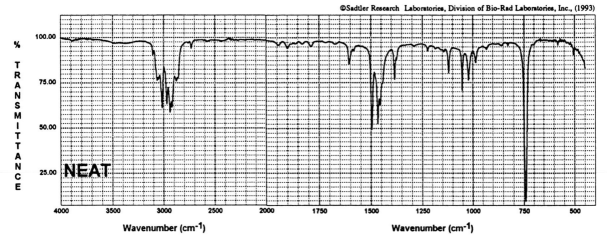

IR unknown spectrum *c.*

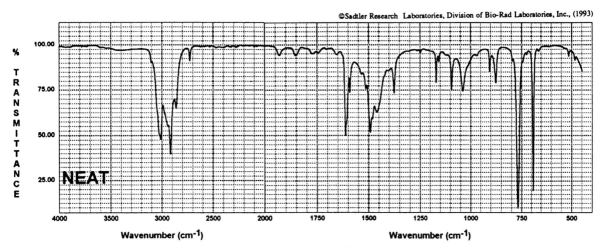

IR unknown spectrum *d.*

6-15. The C—H stretching mode of chloroform (CHCl$_3$), which occurs at 3022 cm^{-1}, is one of the rare exceptions to the 3000-cm^{-1} rule. What is the rule? Suggest an explanation for this exception.

BIBLIOGRAPHY

Aldrich Library of Infrared Spectra, Aldrich Chemical Co., Inc., 940 West Saint Paul Avenue, Milwaukee, WI 53233. 3rd ed., 1981, 12,000 spectra, eight per page, in one volume arranged by chemical type.

Aldrich Library of FT-IR Spectra, Aldrich Chemical Co., Inc., 940 West Saint Paul Avenue, Milwaukee, WI 53233. 1st ed., 1985, 11,000 spectra, four per page, in two volumes arranged by chemical type.

API collection (American Petroleum Institute). About 4500 spectra. M.C.A. collection (Manufacturing Chemists' Association). About 3000 spectra. Chemical Thermo-dynamics Property Center, Texas A&M, University Department of Chemistry, College Station, TX 77843.

Coblentz Society Spectra, Coblentz Society, P.O. Box 9952, Kirkwood, MO 63122. An extensive collection of critically evaluated spectra. Eleven volumes of 1,000 spectra.

D.M.S. System (Documentation of Molecular Spectra). About 15,000 spectra. IFI/Plenum Data Corp., 227 West 17th Street, New York, 10011.

GRASSELLI, J. G.; RITCHEY, W. M., Eds., *Atlas of Spectral Data and Physical Constants for Organic Compounds*; CRC Press: 2000 Corporate Blvd. NW, Boca Raton, FL 33431. 2nd ed., 1975.

Sadtler Library. About 140,000 spectra of single compounds; about 12,000 spectra of commercial products. Sadtler Research Labs, 3316 Spring Garden Street, Philadelphia, PA 19014.

7

Introduction to Nuclear Magnetic Resonance Spectroscopy

TECHNIQUE 11

NUCLEAR SPIN

Nuclear spin is an energy property intrinsic to a nucleus and analogous to the electron spin that plays such an important role in determining electron configurations. Nuclear spin values are quantized, as are electron spins, and are represented by I, the nuclear spin quantum number. Nuclear spin quantum numbers range from 0 through $\frac{7}{2}$ in increments of $\frac{1}{2}$. The nuclei of greatest interest to organic chemists, the ^1H, ^{13}C, ^{19}F, and ^{31}P nuclei, have spins of $\frac{1}{2}$; the ^{12}C, ^{16}O, and ^{32}S nuclei have spins of 0 (and thus cannot be observed by nuclear magnetic resonance spectroscopy, NMR); the ^2H (deuterium, D) and ^{14}N nuclei have spins of 1. Since any spinning charged particle (or body) produces a magnetic moment, a nucleus with a nonzero spin quantum number has a magnetic moment, μ.

Nuclear spin values are quantized because the nuclear angular momentum, and thus the nuclear magnetic moment, is quantized. When placed in an external magnetic field, nuclei orient their magnetic moments in certain ways with respect to the magnetic field, which is assumed to be aligned with the z axis of a Cartesian coordinate system. These orientations are referred to as the z components of the nuclear magnetic moment, μ_z. For a nucleus with a spin of $\frac{1}{2}$, μ_z may be $+\frac{1}{2}$ or $-\frac{1}{2}$. In general, for a nucleus of spin I, the μ_z takes quantized values from $[-I, -I + 1, \ldots, I - 1, I]$; or $(2I + 1)$ different values in all. For this discussion we will limit ourselves to nuclei with spin $\frac{1}{2}$, since this is easier to describe, and since most nuclei of interest in organic chemistry are of spin $\frac{1}{2}$.

When placed in a static magnetic field of strength H_0, the magnetic moment, μ_z, of the spinning nucleus precesses about the magnetic field at a frequency, ν, such that $\nu = \gamma H_0/2\pi$ where H_0 is the strength of the applied magnetic field, and γ is a characteristic property of the nucleus known as the gyromagnetic ratio. When a nucleus of spin $\frac{1}{2}$ placed in a magnetic field, the energies of the $\mu_z = +\frac{1}{2}$ and $-\frac{1}{2}$ states are separated, since in one spin state the nuclear magnetic moment is aligned with the applied magnetic field, and in the other spin state the nuclear magnetic moment is opposed to the applied magnetic field.

The amount of separation of the two energy states, ΔE, is proportional to the magnetic field, and is given by the following expression:

$$\Delta E = \frac{h\gamma H_0}{2\pi} = h\nu$$

When nuclei in the magnetic field are exposed to radiation of the proper frequency, transitions between the two energy states are stimulated, and the nucleus is said to be in *resonance*, or to *resonate*. This transition occurs when the frequency and the energy difference are related by the Planck relation, $\Delta E = h\nu$, and thus the sample will absorb energy of frequency ν. The study of these energy changes is known as nuclear magnetic resonance, or NMR, spectroscopy.

INSTRUMENTATION

In an NMR spectrometer, the magnetic field is provided by a large permanent magnet, electromagnet, or superconducting electromagnet. Commercially available NMR spectrometers have magnets with field strengths that range from 1.4 to 16.3 tesla (the earth's magnetic field at its surface is roughly 5×10^{-5} tesla), and thus operate at frequencies from 60 to 700 MHz for protons. In general, most spectrometers with an operating frequency above 100 MHz use a superconducting electromagnet.

Traditionally, NMR spectra were acquired either by holding the applied magnetic field constant and sweeping the radio frequency (rf), or by holding the rf constant and sweeping the applied magnetic field. Energy absorption by the sample was detected, and the result was the NMR spectrum, a plot of intensity (of energy absorption) versus frequency (or field). This instrumental technique is referred to as continuous wave, or CW, spectroscopy (Fig. 7.1) Over

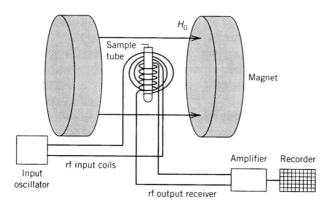

Figure 7.1 Schematic of NMR spectrometer. *(Reprinted with permission of John Wiley & Sons, New York.)*

the last 20 years, however, it has been commonly replaced by pulsed, or Fourier transform (FT), NMR spectroscopy. Among many other benefits, FT-NMR spectroscopy allows very rapid acquisition of spectral data, which permits analysis of small samples and rare nuclei, such as ^{13}C.

The basic principles of FT-NMR spectroscopy can be qualitatively explained as follows. Take, for example, an NMR spectrum that contains a single peak at a given frequency. The graph of this spectrum (Fig. 7.2) is a plot of intensity versus frequency. The same information can be conveyed by a plot of intensity versus time that shows a cosine wave at the frequency described by the graph of the usual NMR spectrum. This is shown in Figure 7.3, and for a spectrum with a single frequency, this plot of intensity versus time is almost as easy to interpret as the usual NMR spectrum shown in Figure 7.2

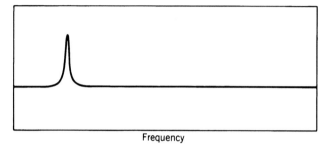

Frequency

Figure 7.2 Intensity versus frequency (usual NMR spectrum).

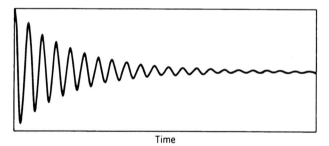

Time

Figure 7.3 Intensity versus time.

Of course, it would be very difficult to determine the frequencies of many superimposed cosine waves from this kind of plot, and it would be at best awkward to interpret a complex NMR spectrum presented in such a fashion (Fig. 7.4) The use of the Fourier transform allows us mathematically to interconvert these time domain (Fig. 7.4) and frequency domain spectra (Fig. 7.5). Fourier transform of the apparently complex spectrum in Figure 7.4 gives the spectrum in Figure 7.5. It is then easy to see that there are actually only three different resonance signals contained in the time domain data of Figure 7.4.

Fourier transform NMR spectra are obtained by applying a short (~1–10 μs), high-powered pulse of rf energy to the sample (Fig. 7.6). This pulse affects all the nuclei to be observed. Before the pulse is applied, the equilibrium net nuclear magnetization is aligned with the applied magnetic field, along the z axis. The coordinate system is presumed to be rotating about the z axis at the frequency of the rf pulse. The pulse, applied down the x axis, applies a torque to the nuclear magnetic moments and rotates them into the xy plane. At this point the pulse is turned off and the nuclear magnetic moments return to their equilibrium alignments. In the process, they precess about the z axis (applied

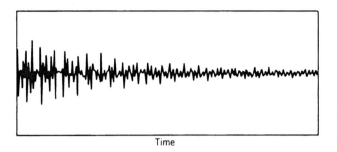

Figure 7.4 Three-signal NMR spectrum: intensity versus time.

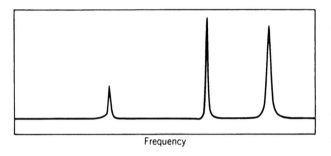

Figure 7.5 Three-signal NMR spectrum: intensity versus frequency.

magnetic field) in the xy plane and induce a current in a detector coil, which can be thought of as being aligned with the y axis. This current varies in a sinusoidal manner, and the observed frequency will be the difference between the resonance frequency of the nuclei and the frequency of the rf pulse.

The detected signal, which is called the free induction decay (FID), is digitized and stored. For small organic molecules in a nonviscous solution, the FID will disappear after a few seconds, which corresponds to the time it takes the nuclei to regain equilibrium alignments after the rf pulse. Thus, an entire ^{1}H NMR spectrum can be obtained in approximately 2 s, in contrast to the 10–15 min usually needed to obtain a CW spectrum. A major advantage of FT-NMR is that many spectra of a sample can be rapidly obtained and added together to increase the signal-to-noise (S/N) ratio. Noise is presumably random about some zero level, so when many spectra are added together, the noise level is reduced, while real signals are reinforced when added. The S/N ratio is proportional to the square root of the number of spectra added together. Thus, one can obtain the ^{1}H spectrum of a 10-μmol sample in a minute or two. With FT-NMR it is possible to obtain spectra of isotopes that are insensitive and/or of low natural abundance, as well as spectra of large biological mole-

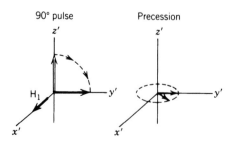

Figure 7.6 Basic pulsed NMR experiment.

cules in dilute solution. By adding a few hundred spectra together, an adequate ^{13}C NMR spectrum of a 100-μmol sample can be obtained in about 20−30 min.

CHEMICAL SHIFT

In a molecule, the magnetic field at a nucleus depends not only on H_0, the field generated by the instrument (the external field), but also on the magnetic fields associated with the electron density near the nucleus. Electrons are influenced by the external field in such a way that their motion generates a small magnetic field that opposes the applied field, and reduces the actual field experienced at the nucleus. This reduction is very small (relative to the external field) and is on the order of 0.001%, 10 ppm, for most protons, and about 200 ppm for ^{13}C nuclei. Reduction of the external field is known as *shielding*, and it gives rise to differences in the energy separation for nuclei in different electronic environments in a molecule. The differences in the energy separation are known as *chemical shifts*.

The magnitude of the chemical shift depends on the nature of the valence and inner electrons of the nucleus and even on electrons that are not directly associated with the nucleus. Chemical shifts are influenced by inductive effects, which reduce the electron density near the nucleus and reduce the shielding. The orientation of the nucleus relative to π electrons also plays an important role in determining the chemical shift. A proton located immediately outside a π-electron system (as in the case of the protons on benzene rings) will be significantly deshielded. In most molecules the chemical shift is determined by a combination of these factors. Chemical shifts are difficult to predict using theoretical principles, but have been well studied and can usually be easily predicted empirically upon comparison to reference data.

In an NMR spectrum, the absorption of rf energy is detected, as in Figure 7.7, where the energy absorption is shown for increasing frequency. In this example we illustrate the case with two different nuclei, A and X. Since A and X are different, they absorb energy at different frequencies while in the same applied magnetic field.

The spectrum would be displayed as in Figure 7.8. The difference in the resonances is the *chemical shift* and is expressed in parts per million (ppm). The

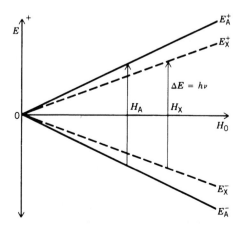

Figure 7.7 The energy splitting for two chemically different protons. The differences between the A energy levels (solid lines) and the X levels (dashed lines) have been amplified for illustrative purposes. At 60 MHz, nucleus A absorbs energy at field H_A and nucleus X absorbs energy at field H_X. Nucleus X is said to be more strongly shielded than A. The resonance for X is said to occur upfield of that for A.

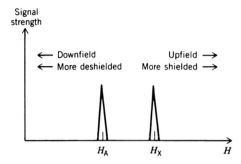

Figure 7.8 The spectrum for the system in Figure 7.7 as it would be displayed. It is conventional to display the spectrum with magnetic field strength increasing to the right so that upfield (and more strongly shielded) is toward the right and downfield (and deshielded) is toward the left.

use of frequency units is cumbersome and is complicated because NMR spectrometers of different magnetic field strengths (and thus operating frequencies) are used. The use of ppm units allows direct comparisons of spectroscopic data obtained on different instruments, when chemical shifts are referenced relative to a reference compound whose chemical shift is arbitrarily defined as 0 ppm. The accepted reference standard for ^1H and ^{13}C NMR in organic solutions is tetramethylsilane (TMS), $(CH_3)_4Si$. The chemical shift relative to TMS is symbolized by δ, which is defined below.

$$\delta = \frac{10^6 \, (\nu - \nu_{TMS})}{\nu_{TMS}}$$

Tetramethylsilane is used as a reference substance for a number of reasons. It is more strongly shielded (Si is more electropositive than C) than most other protons and carbon atoms, and its resonance is thus well removed from other areas of interest in the NMR spectrum. Tetramethylsilane is inert and thus unlikely to react with the compound being analyzed, it is volatile (bp 26 °C) and thus easily removed after a sample has been analyzed, and its 12 identical protons per molecule provide a strong signal per molecule of TMS.

SPIN–SPIN COUPLING

In a molecule with several protons, the exact frequency at which a proton resonates depends not only on the chemical shift of that proton, but also on the spin states of nearby protons. This occurs because the magnetic moments of the nearby protons can either shield or deshield the proton in question from the applied magnetic field, depending on the orientation of the nearby magnetic moments relative to the applied magnetic field. The extent of this perturbation is independent of the applied magnetic field strength. The effect of the spin state of one nucleus on the resonance of another is known as *coupling* or *splitting*.

The spectra resulting from spin–spin coupling depend on the types of nuclei, the distance and geometry between the nuclei, the nature of the bonding, the electronic environment, and the total number of spin states possible. The latter may be illustrated by looking at the spectrum of an imaginary compound that has protons H_A and H_X on adjacent carbons, connected by three bonds: $H_A—C—C—H_X$ (Fig. 7.9). In the first approximation we would expect one resonance for H_A and one resonance for H_X, and the spectrum would resemble that shown in Figure 7.8. In the presence of coupling, the resonance for H_A splits into two signals, one of which corresponds to H_X, having $\mu_z = +\frac{1}{2}$ and

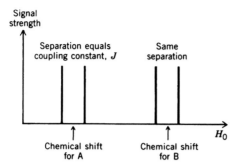

Figure 7.9 Spectrum of two chemically different protons that are coupled.

the other to $\mu_z = -\frac{1}{2}$. The coupling effect is symmetric in that the H_X resonance also splits into two resonances, one for each spin state of H_A. The magnitude of the separation of the H_A pair (a doublet) or the H_X pair (also a doublet) is known as the coupling constant, or J. It is usually expressed in frequency units (Hz), since J is independent of the magnetic field strength.

A simple way to explain this is to consider the effect the two possible spin states of H_X have on the resonance frequency of H_A. The equilibrium population distribution of the two spin states in H_X is very close to 1:1, since ΔE is only about 10^{-3} cal/mol. Since H_X has a magnetic moment, there are then two slightly different magnetic fields at H_A. We thus see two signals for H_A, one for those H_A nuclei adjacent to H_X nuclei with μ_z aligned with the applied magnetic field, and one signal for those H_A nuclei adjacent to H_X nuclei with μ_z aligned opposed to the applied field. Coupling between protons connected by more than three bonds does occur, but its magnitude, J, is usually small and often not directly observed in a usual NMR spectrum.

The splitting becomes more interesting when there are several nuclei of one type. 1,1-Dibromoethane (CH_3CHBr_2) has three equivalent protons in the methyl group and one proton on the C-1 atom. The methyl group exhibits rapid internal rotation so that its three protons are equivalent. The chemical shift for the C-1 proton is 5.86 ppm and that for the methyl protons is 2.47 ppm. Here we can see an example of decreased shielding resulting from the presence of electronegative substituents. Equivalent protons do not couple with one another (this is an important rule in interpreting spectra), but the methyl protons will affect the proton on C-1, and vice versa.

To analyze the splitting pattern, we need to consider the orientations of the nuclear magnetic moments, with respect to the applied magnetic field, for all three methyl protons. Since each of the three protons may have two spin states that are of nearly equal probability, there are $2^3 = 8$ possible combinations of spin states in all for the methyl protons. The net sums of these may have only four different values, as shown in Figure 7.10. The symbol $(+)$ is used to represent $\mu_z = +\frac{1}{2}$ for a single proton and $(-)$ is used to represent $\mu_z = -\frac{1}{2}$. Thus $(+)(+)(-)$ means that protons 1 and 2 have $\mu_z = +\frac{1}{2}$, while proton 3 has $\mu_z = -\frac{1}{2}$.

The number of different μ_z states is $(2N + 1)$, where N is the number of equivalent nuclei (of spin $\frac{1}{2}$). Thus the three methyl protons can generate four slightly different magnetic fields, and the proton on the C-1 of CH_3CHBr_2 sees (in different molecules) four different magnetic fields. Since these different magnetic fields are not of equal probability, but rather are populated in a ratio of 1:3:3:1, the four signals we see for the proton on C-1 when coupled to the methyl group are of intensities 1:3:3:1, and are referred to as a *quartet*. This is shown schematically in Figure 7.11.

Individual μ_z	SUM
(+)(+)(+)	$+\frac{3}{2}$
(+)(+)(−) or (+)(−)(+) or (−)(+)(+)	$+\frac{1}{2}$
(+)(−)(−) or (−)(+)(−) or (−)(−)(+)	$-\frac{1}{2}$
(−)(−)(−)	$-\frac{3}{2}$

Figure 7.10 Possible combinations of spin states for a methyl group.

Since the proton on C-1 has two possible spin states of nearly equal probability, the protons of the methyl group experience two slightly different magnetic fields and are observed in the spectrum as two slightly separated signals of equal intensity, or a *doublet*. The separation between each of the C-1 proton signals is the coupling constant, J, and will equal the J of the methyl signal. The proposed spectrum is shown in Figure 7.11*b*. The coupling constant in this case is about 7 Hz. The 60-MHz NMR spectrum of 1,1-dibromoethane is shown in Figure 7.11*c*.

Note

The net effect of spin–spin coupling is that a proton (or group of equivalent protons) adjacent to N other protons will be observed as a multiplet with $(N + 1)$ lines.

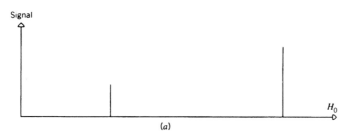

(a)

Figure 7.11a The "stick figure" spectrum without any spin–spin coupling.

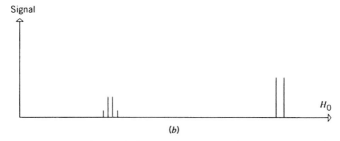

(b)

Figure 7.11b The "stick figure" spectrum with spin–spin coupling.

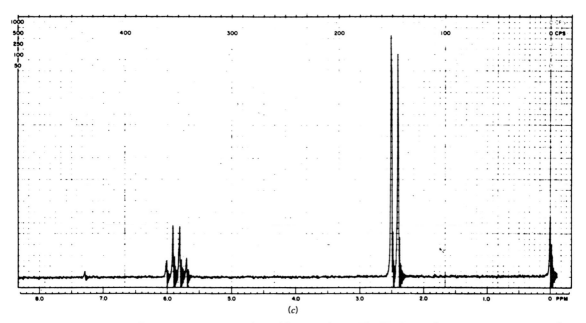

(c)

Figure 7.11c The actual 60-MHz spectrum of 1,1-dibromoethane. The TMS signal at 0 ppm is seen as well as a weak signal at 7.3 ppm, which is not from this molecule.

A proton, or group of equivalent protons, may be coupled to more than one group of nuclei. The spectrum of 1-nitropropane ($CH_3CH_2CH_2NO_2$) is shown in Figure 7.12. The signal from the central methylene (CH_2) group is seen at about 2.0 ppm. Because the methylene group is adjacent to (and thus coupled to) five protons, its signal is a (5 + 1) or six-line multiplet—a *sextet*.

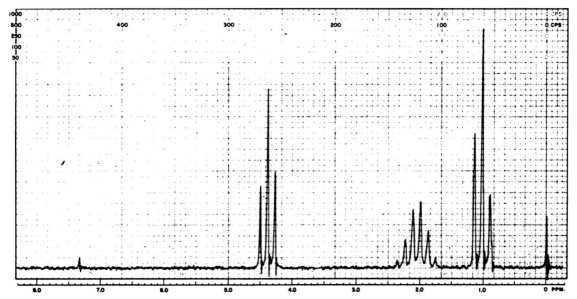

Figure 7.12 The 60-MHz spectrum of 1-nitropropane. (*Courtesy of Varian Associates, Palo Alto, CA.*) Starting from the right, the TMS signal at 0 ppm is seen. Next is a 1:2:1 triplet at 1.03 ppm. This triplet results from the protons on C-3 and their coupling with the two protons on C-2. Next is the sextet centered at 2.07 ppm. This multiplet is from the protons on C-2 and their coupling with the protons on C-1 and C-3. Finally, we have the signal from the protons closest to the nitro group centered at 4.38 ppm. These protons appear as a 1:2:1 triplet due to their coupling with the protons on C-2.

187

Nuclei with spins of 1 or greater exhibit more complex spin–spin coupling, since they can exist in more than two different spin states. For a nucleus coupled to N nuclei of spin I, a multiplet of $2IN$ lines will be observed. Nuclei of spin zero do not couple.

INTENSITIES

The area under an NMR peak is proportional to the number of nuclei giving rise to that signal. The intensity of a resonance is thus best determined by the *integral* of the NMR spectrum over a resonance, or group of resonances. Nuclear magnetic resonance spectrometers can measure the integral, though integration data from an FT spectrometer are less reliable than those from a CW spectrometer. In more complex spectra the intensities are useful as a measure of the number of protons of a given type. For instance, in the above case the integral over both peaks of the methyl group doublet will be three times the integral over the quartet of the proton on C-1. Integration can thus often provide useful information for determining the identity of a compound.

SECOND-ORDER EFFECTS

So far, all of our examples have consisted of first-order spectra. First-order spectra are those multiplets interpretable through elementary coupling analysis, such as that above; second-order spectra are those that are not interpretable in this manner. These highly symmetric and fairly simple first-order spectra are generally observed when the chemical-shift differences (expressed as a frequency) are much greater than the coupling constant. Second-order effects occur when the coupling constants become comparable to or greater than the chemical-shift differences. Thus, spectra obtained on instruments with higher magnetic fields are more likely to be first order, since the frequency differences between given signals increase with increasing magnetic fields. However, the chemical-shift differences (in ppm) remain the same regardless of magnetic field strength.

Second-order effects may be understood in qualitative terms by considering the limiting cases. Let us consider the hypothetical disubstituted ethylene shown in Figure 7.13, where R and M are substituents that might be identical or may have very different effects on the alkenyl protons. In Figure 7.13*a* the spectrum is shown for the situation in which R and M have very different effects. In this case we will observe a first-order spectrum consisting of two doublets. The coupling constant is the separation in the doublets, and the chemical shift of each nucleus is the geometric midpoint of each doublet.

In Figure 7.13*b* groups R and M are identical. H_A and H_B are identical in this case and only a single resonance is observed (coupling between equivalent nuclei is not observed).

In Figure 7.13*c* the difference in the chemical environment of H_A and H_B is very slight. The spectrum shown may be seen as intermediate between the limiting cases in Figure 7.13*a* and Figure 7.13*b*. Note that there is a "leaning in" of the doublets as the central members increase in intensity at the expense of

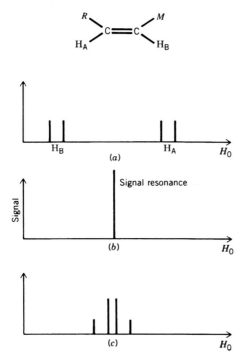

Figure 7.13 Second-order effects. (a) The chemical-shift difference is much larger than the coupling constant, and a first-order spectrum is observed. (b) Protons A and B are equivalent, and a single resonance is observed. (c) The chemical-shift difference is of the same order of magnitude or less than the coupling constant. Note the "leaning in" of the peak intensities in this spectrum relative to that in part *a*.

the outer members. A full continuum of behavior may be expected with cases observed in which the outer members are lost in the noise and the central members take the appearance of a doublet. This would be one example of a class of spectra known as "deceptively simple spectra."

The second-order spectra of systems with more than two protons are difficult to describe even in qualitative terms. Second-order spectra may well display more lines than one would predict from simple coupling theory. Also, in second-order spectra, the coupling constants and the chemical-shift differences may not be obtainable as simple differences in the positions of spectral lines. Thus, spectra obtained at high frequencies (and magnetic fields) are often more useful. As the operating frequency of the instrument is increased, the chemical-shift differences (in frequency terms) increase while the spin–spin coupling remains constant. Thus, the complicating second-order effects are likely to be less noticeable in high-field spectra. The reader is referred to more extensive treatments of NMR for a discussion of second-order cases.

INTERPRETATION OF ^{1}H NMR SPECTRA

The first issues that must be addressed are molecular symmetry and the magnetic equivalence or nonequivalence of protons or other functional groups. Even if two protons, or groups, are chemically equivalent, they may or may not be magnetically equivalent. Although molecular symmetry can often simplify NMR spectra, one must be able to discern which protons or groups are equivalent by symmetry. The two most useful symmetry properties (or symmetry operators) are the plane of symmetry and the axis of symmetry.

A plane of symmetry is simply a mirror plane such that one half of the molecule is the mirror image of the other half, as in *meso*-pentane-2,4-diol:

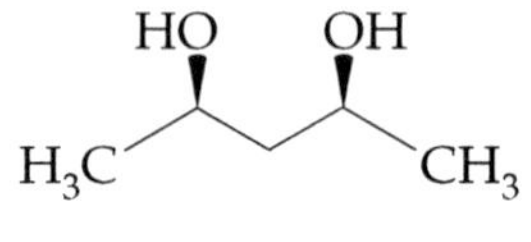

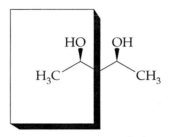

meso-Pentane-2,4-diol

The methyl groups are identical by symmetry, and one would expect this stereoisomer to show one methyl doublet in its ^{1}H NMR spectrum and one signal for a methyl group in its ^{13}C spectrum.

Consider the other diastereomer of pentane-2,4-diol, the chiral *d,l* isomer. This isomer has an axis of symmetry. If the molecule is rotated 180° about an axis in the plane of the paper passing through the central carbon, the molecule can be converted into itself:

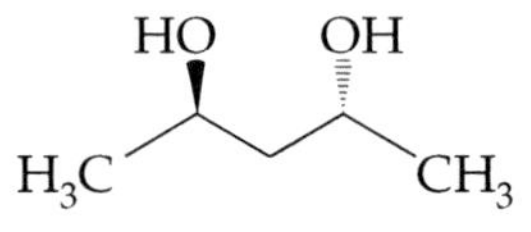

d,l-Pentane-2,4-diol

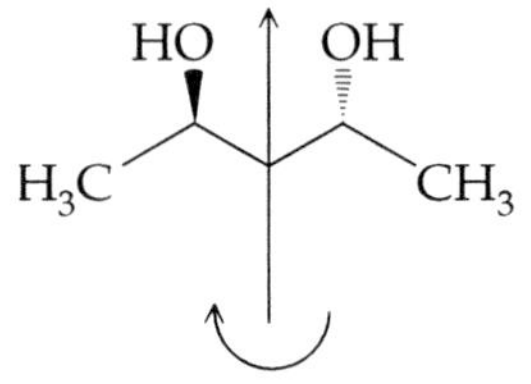

Here, too, the methyl groups are identical by symmetry, and one would expect this stereoisomer to show one methyl doublet in its ^{1}H NMR spectrum and one signal for a methyl group in its ^{13}C spectrum.

It is, however, relatively simple to use NMR spectroscopy to distinguish between these stereoisomers. To do this, look at the two methylene protons on the central carbon, C-3, of each isomer:

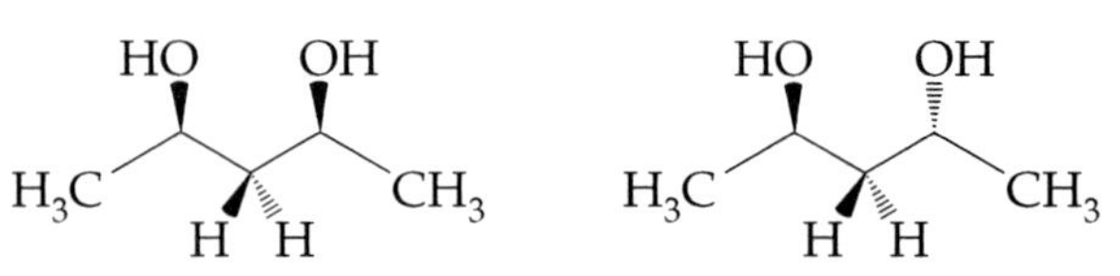

The plane of symmetry in the meso isomer bisects each of the two protons. In the *d,l* isomer, the axis of symmetry interconverts the two protons. Thus, in the *d,l* isomer, the two methylene protons are equivalent by symmetry, but they are not equivalent in the meso isomer. This can also be seen by inspecting the molecule. On the left, one H is syn to both —OH groups and the other is anti to both —OH groups. On the right, each H atom is syn to one —OH group and anti to the other.

The more rigorous way to determine equivalence or nonequivalence is to determine whether the two protons (or groups) are homotopic (identical), diastereotopic, or enantiotopic. To compare two protons, we use the usual Cahn–Ingold–Prelog system for the nomenclature of stereoisomers. We artificially distinguish the relative priority of two protons by a method such as drawing them in different colors or pretending that one is deuterium (as long as the molecule does not contain D). We draw the two possibilities (i.e., the first H as D and then the second H as D) and then determine the stereochemical relationship between the two:

If the two are identical, the two protons are identical, or *homotopic*. If the two structures are diastereomers, the two protons are *diastereotopic*, and if the two structures are enantiomers, the two protons are *enantiotopic*. Diastereotopic protons, or groups, will be magnetically nonequivalent. Enantiotopic protons, or groups, will be magnetically equivalent only in an achiral environment and may appear nonequivalent in a chiral environment, such as a chiral solvent or in a biological sample. Homotopic protons may or may not be magnetically equivalent. Of course, it is possible for magnetically nonequivalent signals to be so close to one another in the NMR spectrum as to overlap (accidentally degenerate).

Homotopic protons may be magnetically nonequivalent if the two protons have different coupling constants to the same third proton. The most common example of this occurs in para-substituted benzenes:

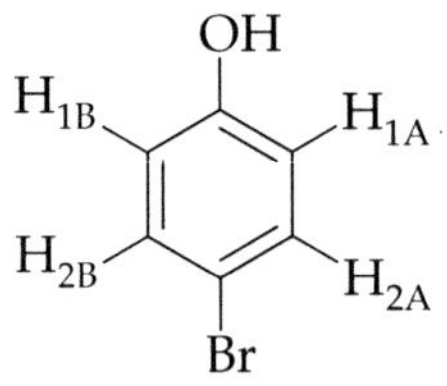

OH

H_{1B} H_{1A}

H_{2B} H_{2A}

Br

By symmetry, H_{1A} and H_{1B} are equivalent. These protons are not, however, magnetically equivalent because H_{1A} and H_{1B} have different coupling constants to, for example, H_{2A}, and the spectrum of this molecule may well be more complex than one would at first expect.

The equivalence or nonequivalence of functional groups, as well as protons, can easily be determined. The ^1H NMR spectrum of menthol shows three methyl doublets, since the two methyls in the isopropyl group are diastereotopic. The ^{13}C spectrum of menthol shows three distinct resonances for the three different methyl groups:

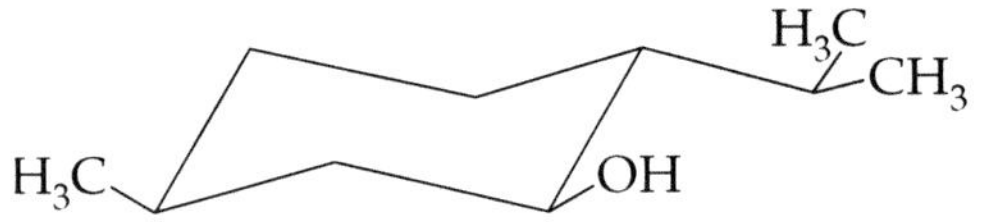

H_3C CH_3

H_3C OH

Menthol

^1H CHEMICAL SHIFTS

Figure 7.14 summarizes the chemical shifts of protons in a large range of chemical environments. It is, however, a bit dangerous to use figures such as this one without understanding some of the factors that underlie shielding and the chemical shift. To give some flavor of the factors that determine chemical shifts and the range of values observed, we will briefly examine chemical shifts in methyl groups and chemical shifts for protons on sp^2 carbon atoms.

Methyl groups bonded to an sp^3 carbon generally have chemical shifts in the range 0.8–2.1 ppm as long as there is no more than one electron-withdrawing group attached to the carbon. The shifts generally increase as the strength of the electron withdrawing group increases, or as more electron-withdrawing groups are added. Groups that inductively withdraw electrons reduce the electron density near the methyl group protons. This results in less shielding and a downfield shift of the methyl resonance. This effect is clearly seen in the spectra of 1,1-dibromoethane and 1-nitropropane (Figure 7.11c and Figure 7.12), respectively. The chemical shifts for methyl groups bonded to sp^2 carbon atoms fall in the range 1.6–2.7 ppm.

In the case of a proton bonded to an sp^2 carbon, the location of the proton relative to the π cloud plays an important role in determining the chemical shift. In unconjugated alkenes chemical shifts fall in the range 5–6 ppm. Where more than one proton is bonded to an alkene, complex second-order spectra can be expected at low operating frequencies since the coupling constants are usually fairly large relative to the difference in resonance frequencies.

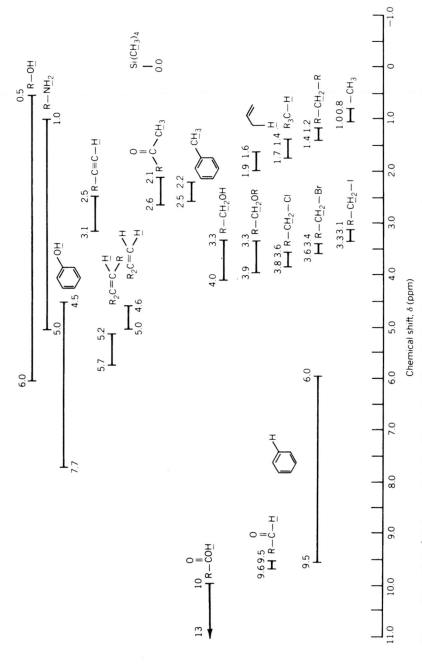

Figure 7.14 NMR ^{1}H chemical shifts. *(From Zubrick, J. W. The Organic Lab Survival Manual, 4th ed.; Wiley: New York, 1997. Reprinted by permission of John Wiley & Sons, New York.)*

In aldehydes, RCHO, the increased electronegativity of the oxygen increases the deshielding and the chemical shift falls in the range 9–10.5 ppm.

The chemical shift in an aromatic system is generally greater than that for alkenes. For example, the chemical shift of benzene is 7.37 ppm, which is substantially greater than the 5.6 ppm for the alkenyl protons of cyclohexene. Much of this difference results from the "ring current" effect and the orientation of the proton relative to the aromatic π electrons. If the ring substituents are not strongly electron withdrawing or electron donating, such as alkyl groups, the chemical shift for ring protons will not be shifted greatly from that of benzene itself. Furthermore, these substituents generate only small chemical-shift differences among the ring protons. Thus, the 60-MHz spectra for toluene (methylbenzene) appears to have a single resonance in the aromatic region at about 7.1 ppm. If, on the other hand, the substituents are electron withdrawing, the ortho and para ring protons will be somewhat deshielded relative to benzene. Pi-electron-donating substituents, such as a methoxy group, will increase the shielding of groups ortho and para to it.

SPIN–SPIN COUPLING

Coupling information is the primary reason that ^1H NMR is such a powerful tool for organic structure determination. Since coupling information is transmitted through bonds, coupling provides information about nearby protons and can often be used to deduce stereochemistry.

The sign of the coupling constant (usually symbolized as J) may be positive or negative. However, first-order spectra are not sensitive to the sign of the coupling constant. In second-order cases, the sign of J may be determined by a detailed analysis of the spectrum, though the sign of J is generally of little value for organic structure determination.

Geminal coupling

Nonequivalent protons attached to the same carbon (geminal protons) will couple with one another. These coupling constants tend to be large (>10 Hz) for sp^3 carbon atoms and small (<4 Hz) for sp^2 carbon atoms. Geminal coupling constants tend to decrease with decreasing ring size, because of hybridization changes at carbon, and with the increasing electronegativity of the substituents on a given methylene group.

Vicinal coupling

Vicinal coupling describes the coupling over three bonds observed between protons attached to two bonded carbon atoms, H—C—C—H. Vicinal coupling constants (J values) can range from near 0 to greater than 15 Hz, depending on the stereochemical relationship (dihedral angle) between the coupled protons, the hybridization of the carbon atoms, and the electronegativity of other substituents. For vicinal protons on sp^3 carbon atoms, the coupling constant is related to the dihedral angle and is expressed graphically by the Karplus curve (Fig. 7.15).

Though the magnitude of vicinal coupling is very sensitive to the angle of rotation about the central bond, in many simple cases nearly all coupling

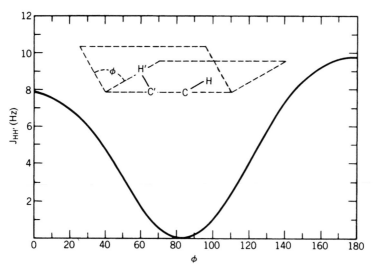

Figure 7.15 The vicinal Karplus correlation showing the relationship between dihedral angle and coupling constants for vicinal protons.

constants are equal. This situation is often the case if internal rotation about a C—C single bond can occur on a time scale that is very short relative to the NMR time scale, such as in acyclic systems. In these cases, the effect of internal rotation is completely blurred as far as NMR is concerned, and only an average coupling constant is observed. Vicinal coupling constants in freely rotating alkyl groups are usually observed in the 6.5- to 8-Hz range.

When the central C—C bond between two coupled protons is a double bond, rotation is restricted and separate coupling constants for cis and trans protons may be observed. Cis coupling constants fall in the range 5–12 Hz, whereas trans coupling constants range from 12 to 20 Hz. As a result of these large coupling constants, second-order effects are often observed in substituted alkenes in instruments of lower field strengths.

When rotation about carbon–carbon single bonds is restricted, or when stereochemistry dictates significant conformational preferences, nonaveraged coupling constants may be observed that can complicate the appearance of the NMR spectrum. For example, two diastereotopic protons of a methylene (CH_2) group may well each couple to a given third proton with different coupling constants. The familiar coupling explained at the elementary level suggests that if one proton is adjacent to, for example, two others, the NMR signal of the first proton will be a triplet. This simplification will be true only if the two coupling constants are identical. A triplet is merely a doublet of doublets with equal coupling constants, which gives rise to the familiar triplet with intensities of 1:2:1 (Fig. 7.16a). A doublet of doublets, on the other hand, gives rise to a four-line multiplet with peaks of roughly equal intensity (Figure 7.16b).

Long-range coupling

Longer range coupling involving four or more bonds is common in allylic systems and in aromatic rings and other conjugated π systems. These coupling constants are generally smaller than the values considered above (i.e., <3 Hz).

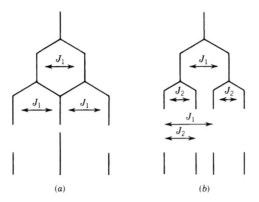

(a)   (b)

Figure 7.16 (*a*) Triplet equals doublet of doublets with equal *J* values; (*b*) doublet of doublets.

EXAMPLES OF COMPLEX, YET FIRST-ORDER, COUPLING

Ethyl vinyl ether

The coupling constants of even a seemingly complex multiplet can be discerned in a relatively simple manner. First, the total width (outside peak to outside peak) of a first-order multiplet is equal to the sum of all the coupling constants, keeping in mind that, for example, a triplet of $J = 7$ Hz is really a doublet of doublets with both J values equal to 7 Hz. The expansion of the proton spectrum of ethyl vinyl ether is presented as an example in Figure 7.17. Integration data are displayed between the spectrum and the horizontal axis in Figure 7.17*a*.

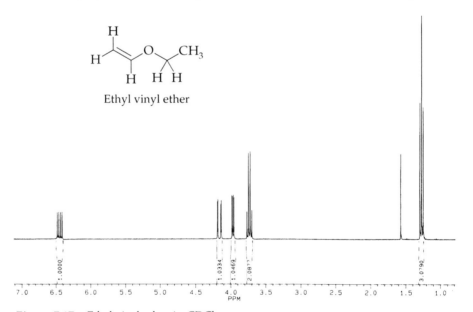

Ethyl vinyl ether

Figure 7.17a Ethyl vinyl ether in CDCl$_3$.

Consider the multiplet centered at 6.45 ppm (Fig. 7.17*b*). By measuring the distance (in Hz) from either outside line to the next inner line, which is 6.8 Hz, the first coupling constant is determined. Then, by measuring from the outside line to the second line in, the second coupling constant is found to be 14.4 Hz. We know that this is the last coupling constant to be found for sev-

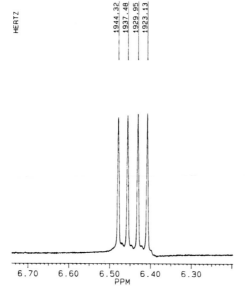

HERTZ

1944.32
1937.48
1929.95
1923.13

6.70   6.60   6.50   6.40   6.30
PPM

Figure 7.17b Ethyl vinyl ether in CDCl₃ (expansion).

eral reasons. First, if we measure from the outside line to the third line in, we get a value, 21.2 Hz, which is equal to the sum of the previously determined coupling constants. Second, the width of the multiplet (the same measurement in this simple case) is equal to our two coupling constants. Thus, the NMR signal at 6.45 ppm is a doublet of doublets with J = 14.4 and 6.8 Hz.

The two doublets of doublets at 4.15 and 3.96 ppm (Fig. 7.17c) must be coupled to one another because they both have the coupling constant of 1.9 Hz in common. This geminal coupling constant is typical of the terminal methylene of an alkene.

Since the proton at 3.96 is coupled to the proton at 6.45 ppm by J = 6.8 Hz, and the proton at 4.15 ppm is coupled to the one at 6.45 ppm by J = 14.4 Hz, the proton at 3.96 ppm must be cis, and the proton at 4.15 ppm must be trans, to the alkene proton at 6.45 ppm.

The simple coupling observed for the ethyl group in ethyl vinyl ether can be readily assigned. The triplet at about 1.25 ppm, which integrates for three protons, is due to the methyl group; it is a triplet because the equivalent protons of the methyl group are coupled to the two protons on the adjacent carbon with equal coupling constants. The O—CH₂ protons are observed in the NMR spectrum as the quartet at about 3.75 ppm; they are a quartet because they are coupled equally to the three equivalent protons of the methyl group:

1.25 ppm

4.15 ⟶ H_A

H_B

H_C

3.96   3.73 ppm

6.45

Chemical shifts

Ethyl vinyl ether

1.9 Hz   H_A   14.4 Hz

H_B

H_C

6.8 Hz

Coupling constants

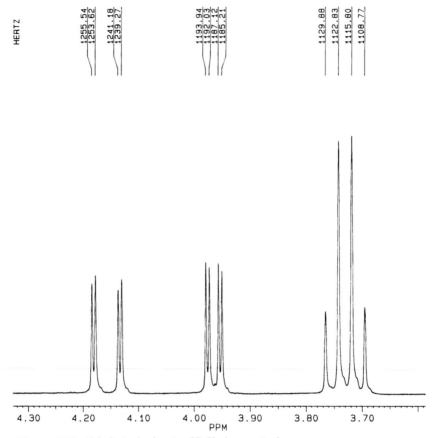

Figure 7.17c Ethyl vinyl ether in CDCl$_3$ (expansion).

Allyl acetate

For a more complex example, refer to the ^{1}H NMR spectrum of allyl acetate (the NMR signal for the methyl group has been omitted) in Figure 7.18*a*.

Protons A, B, and C are all chemically distinct, and the two protons labeled D are equivalent to one another by symmetry (the plane of the paper). The multiplet at 4.58 ppm (Fig. 7.18*d*) corresponds to H$_D$ and is a doublet of triplets. The coupling constant for the triplet is 1.4 Hz, and the coupling constant for the doublet is 5.8 Hz, which can be measured between any two corresponding peaks in the two triplets.

The four quartets around 5.3 ppm (Fig. 7.18c) are actually two doublets of quartets at 5.32 and 5.24 ppm and correspond to H_A and H_B in the structure above. At 5.32 ppm, the multiplet is a doublet of quartets, $J = 17.2, 1.5$ Hz. At 5.24 ppm, we have another doublet of quartets, $J = 10.4, 1.3$ Hz. We see quartets because the long-range allylic coupling to the two H_D signals gives a triplet that has a coupling constant J that is approximately equal to the geminal coupling constant ($\sim$1.4 Hz) between H_A and H_B. Since NMR line widths are naturally several tenths of a hertz, it is not possible to distinguish between coupling constants such as these that differ only by 0.2 Hz. We can unambiguously distinguish H_A and H_B by the magnitudes of their coupling constants to H_C, which are 17.2 and 10.4 Hz. Since trans coupling constants are larger than cis coupling constants, H_A must have the 17.2-Hz coupling constant to H_C and is

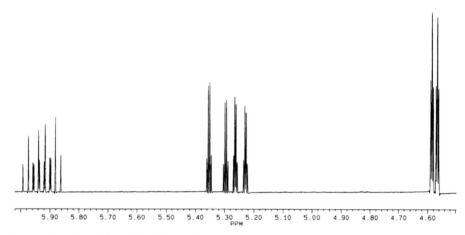

Figure 7.18a Allyl acetate in CDCl$_3$.

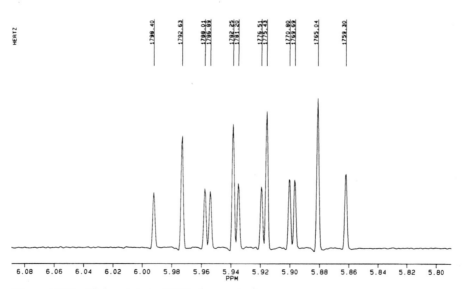

Figure 7.18b Allyl acetate in CDCl$_3$ (expansion).

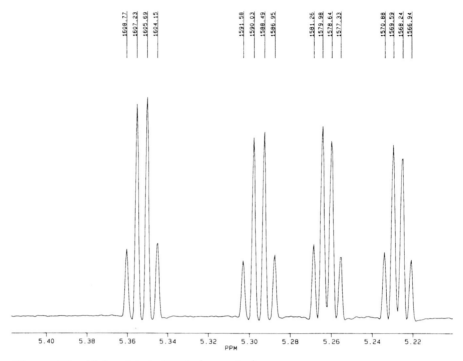

Figure 7.18c Allyl acetate in CDCl₃ (expansion).

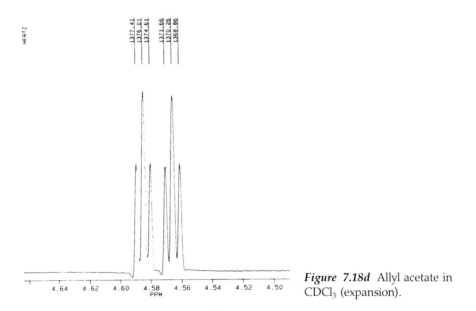

Figure 7.18d Allyl acetate in CDCl₃ (expansion).

thus assigned to the signal centered at 5.32 ppm. Since H_B is coupled to H_C by J = 10.4 Hz, it is assigned to the signal centered at 5.24 ppm.

Finally, we already know what the multiplet for H_C should look like, since we know all of its coupling constants. It is coupled to the two H_D protons with a coupling constant of 5.8 Hz, to H_A with J = 17.2 Hz, and to H_B with J = 10.4 Hz. The multiplet for H_C at 5.93 ppm should be, therefore, a doublet of doublets of triplets with J = 17.2, 10.4, and 5.8 Hz, respectively. There should be $2 \times 2 \times 3 = 12$ lines, and the width should be $17.2 + 10.4 + (2 \times 5.8) = 39.2$ Hz. There are indeed 12 lines (Fig. 7.18b), and the distance between the outside peaks is 39.1 Hz, which is a perfectly reasonable deviation from the ideal:

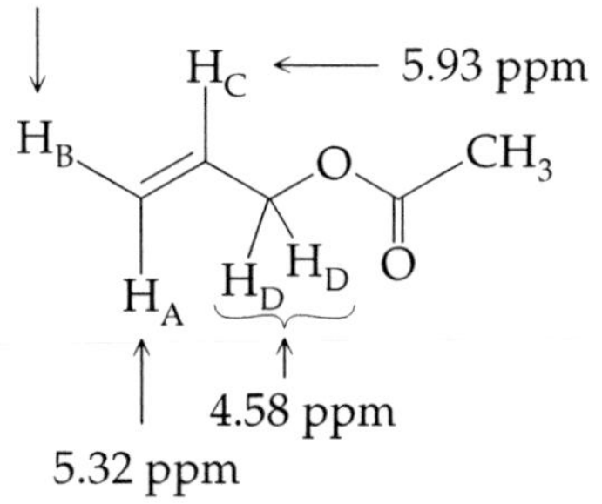

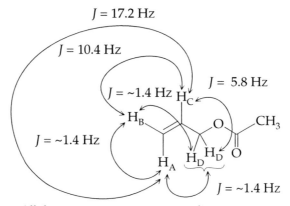

Allyl acetate, chemical shift assignments

Allyl acetate, proton–proton coupling constants

4-t-Butylcyclohexanol

Since cyclohexane rings are often held in at most two potential conformations (both chairs), coupling constants may allow the determination of relative stereochemistry. On cyclohexane rings in chair conformations, the axial–axial coupling constants for vicinal protons (180° dihedral angle) are on the order of 9–12 Hz. Equatorial–equatorial and equatorial–axial coupling constants (60° dihedral angles) are on the order of 2–4 Hz. Thus it is often a relatively simple matter to determine stereochemical relationships on a six-membered ring using NMR spectroscopy.

Take, for example, the two diastereomers of 4-*tert*-butylcyclohexanol:

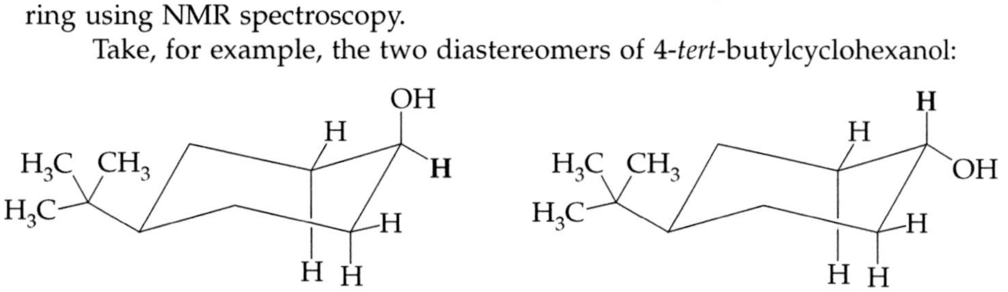

Cis   Trans

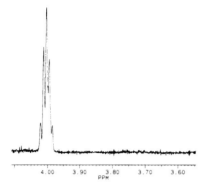

Figure 7.19 *cis*-4-*tert*-Butylcyclohexanol.

We know that the very large *tert*-butyl group will effectively always be equatorial. By examining the coupling to the methine proton of the alcohol, it is simple to determine whether that proton is axial or equatorial. It is also possible to distinguish between these stereoisomers by using chemical-shift information (in one isomer the alcohol methine is seen at ~3.52 ppm and in the other at ~4.04 ppm), but use of coupling information provides a far more definitive and unambiguous determination of stereochemistry.

The alcohol methine proton in the cis isomer is equatorial and thus has a 60° dihedral angle to all four adjacent protons that give rise to a pentet (which is really a doublet of doublets of doublets of doublets with equal coupling constants) with a coupling constant $J = $ ~3 Hz, which is seen in Figure 7.19.

The alcohol methine in the trans isomer is axial and thus has a 180° dihedral angle to each of the two adjacent axial protons and a 60° dihedral angle to each of the two adjacent equatorial protons. This arrangement gives rise to a triplet of triplets with $J = $ ~13 and 3 Hz, which is shown in Figure 7.20.

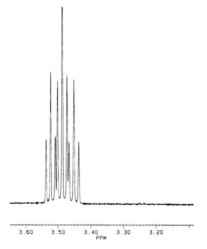

Figure 7.20 *trans*-4-*tert*-Butylcyclohexanol.

¹³C NMR SPECTROSCOPY

With the advent of Fourier transform (FT) NMR spectrometers, ^{13}C NMR spectroscopy is now available as a simple and routine tool for the structure determination of organic molecules. Since ^{13}C is of low natural abundance (1.1%), addition of many spectra is required to obtain acceptable signal-to-noise (S/N) levels. With modern spectrometers, ^{13}C spectra can often be acquired simply by issuing software commands; in some instruments a different probe is inserted into the magnet. Since ^{13}C resonates at roughly 25% of the proton operating frequency of a spectrometer system, an instrument that acquires ^1H spectra at 300 MHz will be reset to about 75 MHz for ^{13}C work.

Generally, ^{13}C NMR spectra are acquired while the entire ^1H frequency range is irradiated by a second rf coil inside the probe assembly. These spectra are referred to as broadband-decoupled ^{13}C spectra and they do not show the effect of spin–spin coupling to ^1H nuclei. Such decoupling is done because ^1H–^{13}C coupling constants can be quite large (a few hundred Hz) relative to chemical-shift differences, which leads to multiplets split over a large portion of the spectrum and subsequent confusion (Fig. 7.21). It is often simpler to see a single line for each distinct carbon atom in a molecule. Furthermore, irradiation of the ^1H spectrum results in signal enhancement of the ^{13}C signals of the attached carbon atoms. This enhancement is the nuclear Overhauser effect (NOE).

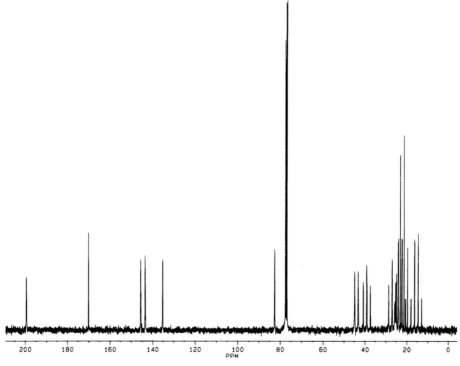

Figure 7.21 Fully ^1H-coupled ^{13}C NMR spectrum of 5-(1-acetoxy-1-methylethyl)-2-methyl-2-cyclohexenone in CDCl$_3$.

The ^{13}C NMR chemical shifts follow the same rough trends as seen in ^{1}H chemical shifts. ^{13}C chemical shifts, however, are not nearly as amenable to prediction based on the electronegativity of substituents as are ^{1}H chemical shifts. The ^{13}C chemical shifts are, in general, less sensitive to substituent electronegativities, and are far more sensitive to steric effects than are ^{1}H chemical shifts. A brief listing of approximate ^{13}C chemical shifts is provided in Table 7.1; a more extensive and thorough listing is available in the Silverstein et al. reference (Bibliography). As in ^{1}H NMR spectroscopy, TMS ($Si(CH_3)_4$) is used as the internal reference and the chemical shift of TMS is defined as zero. Except for functional groups such as acetals and ketals, sp^3-hybridized carbon atoms appear upfield (to the right) of 100 ppm, and sp^2-hybridized carbon atoms appear downfield of 100 ppm. Common carbonyl-containing functional groups appear downfield of 160 ppm. Aldehydes and ketones appear at 195–220 ppm; esters, amides, anhydrides, and carboxylic acids appear at 165–180 ppm.

Typical ^{13}C NMR spectroscopy provides an NMR spectrum that is not amenable to integration because of the NOE and insufficient relaxation delays. Therefore, the number of carbon atoms giving rise to a given signal cannot generally be determined by these techniques. It is possible to obtain ^{13}C NMR spectra that can be accurately integrated (inverse-gated decoupling), but this experiment requires a great deal of acquisition time to achieve adequate signal-to-noise levels.

Information about C—H coupling can be readily obtained, however. Fully coupled ^{13}C NMR spectra are not very useful for structure determination because C—H couplings are large (~120–270 Hz, depending mainly on the hy-

TABLE 7.1 *Approximate ^{13}C NMR Chemical Shifts*

FUNCTIONAL GROUP	CARBON[a]	CHEMICAL SHIFT/ δ (ppm)
Alkyl carbon atoms		~5–45
	1° R—CH$_3$	~5–30
	2° R—CH$_2$—R'	~15–35
	3° R—CHR'R"	~20–40
	4° RCR'R"R'''	~25–45
Alkenyl carbon atoms		~110–150
	H$_2$C=C	~100–125
	HRC=C	~125–145
	RR' C=C	~130–150
Aromatic carbon atoms		~120–160
Alkynyl carbon atoms	C≡C	~65–90
Nitriles	R—C≡N	~115–125
Alcohols and ethers	C—OH(R)	~50–75
	C—O (epoxides)	~35–55
Amines	C—N	~30–55
Alkyl halides	C—X	~0–75
Carbonyl groups	C=O	~165–220
Ketones, aldehydes	RCOR', RCHO	195–220
Carboxylic acids, esters	RCO$_2$H, RCO$_2$R'	165–180
Amides, anhydrides	RCON, RCO$_2$OCR'	160–175

[a]R = alkyl group.

bridization at carbon) and multiplets tend to overlap. Furthermore, when the hydrogen atoms are not irradiated, there is no NOE, and the signal-to-noise ratio suffers significantly. The most common use for coupling information is to determine the number of protons attached to a given carbon atom. This can be done in a variety of ways, some of which do not actually display the carbon signals as multiplets due to coupling to attached protons.

Single-frequency off-resonance decoupling (SFORD) is a useful technique for determining the number of hydrogen atoms attached to a given carbon. The decoupler is tuned off to one side of the proton spectrum and the sample is irradiated at a single frequency giving rise to ^{13}C spectra that show C—H couplings as a fraction of their actual values and that show a partial NOE. The apparent C—H coupling is dependent on both the actual coupling constant and the difference between the decoupler frequency and the resonance frequency of the hydrogen in question. The major disadvantage of SFORD is its low signal-to-noise ratio, which is due to two factors. First, there is only a partial NOE. Second, when NMR signals are split into multiplets, the signal intensity becomes distributed among several peaks. Thus, SFORD spectra require significantly more spectral acquisitions than do fully decoupled ^{13}C NMR spectra, and to some extent have been replaced with distortionless enhancement by polarization transfer (DEPT) spectra.

DEPT ^{13}C NMR spectroscopy provides a rapid way of determining the number of hydrogen atoms attached to a given carbon atom. DEPT spectra result from a multiple-pulse sequence that terminates in a "read pulse," which can be varied according to the spectrum desired. In DEPT spectra, all peaks are singlets; quaternary carbon atoms (without attached hydrogen atoms) are not seen in any DEPT spectra. In DEPT-135° spectra, CH and CH$_3$ groups appear as singlets of positive intensity, and CH$_2$ groups appear as negative peaks. The DEPT-90° spectra show only CH groups, and thus allow CH$_3$ and CH groups to be distinguished. In combination with a routine fully decoupled spectrum, DEPT spectra allow unambiguous assignment of the number of hydrogen atoms attached to each carbon. In practice, such spectral editing techniques are not perfect, and small residual peaks are often seen where, in principle, there should be none; these are usually small enough to be readily distinguished from the "real" peaks.

The fully coupled ^{13}C NMR spectrum of the acetoxy-enone (**I**) is shown in Figure 7.21, the broadband decoupled spectrum in Figure 7.22, and the SFORD spectrum in Figure 7.23. The DEPT-135° spectrum is shown in Figure 7.24, and the DEPT-90° spectrum in Figure 7.25. The 1:1:1 triplet centered at 77 ppm is due to the solvent, CDCl$_3$.

Interpretation and assignment of the ^{13}C NMR spectrum are much easier when we unambiguously know how many protons are attached to each carbon.

I

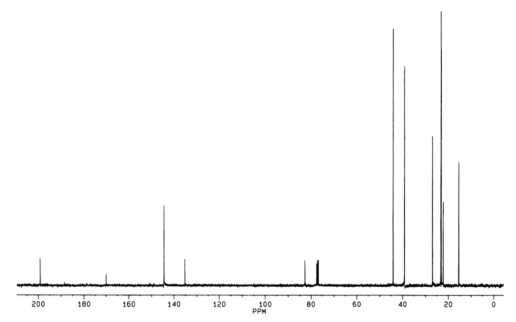

Figure 7.22 Broadband ^{1}H-decoupled ^{13}C NMR spectrum of 5-(1-acetoxy-1-methylethyl)-2-methyl-2-cyclohexenone in CDCl$_{3}$.

The ^{13}C NMR is often better than ^{1}H NMR for distinguishing functional groups because typical ^{13}C chemical shifts are in the range 0–200 ppm relative to TMS, as compared to 0–10 ppm for proton chemical shifts. Coupling between adjacent ^{13}C nuclei is not observed (except in isotopically enriched samples) because the probability of having two rare isotopes adjacent to one another is very small. Because of the absence of coupling, ^{13}C spectra are less complex

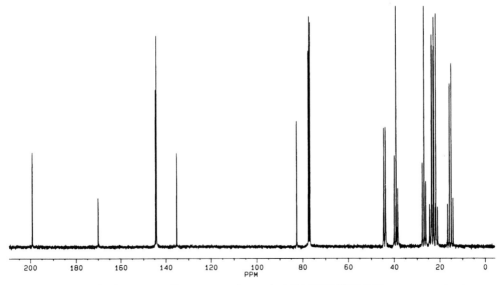

Figure 7.23 Single-frequency off-resonance decoupled (SFORD) ^{13}C NMR spectrum of 5-(1-ace-toxy-1-methylethyl)-2-methyl-2-cyclohexenone in CDCl$_{3}$.

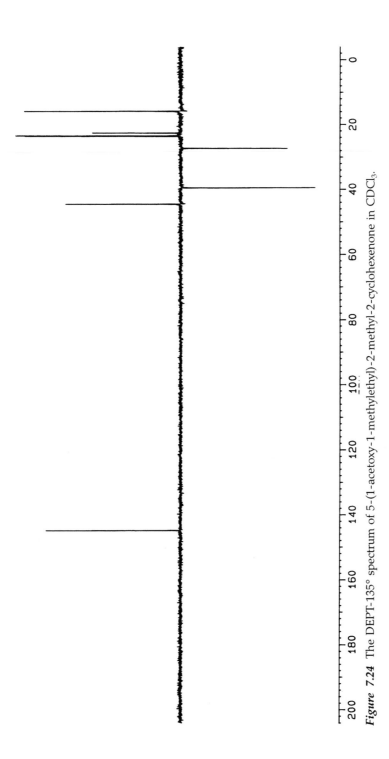

Figure 7.24 The DEPT-135° spectrum of 5-(1-acetoxy-1-methylethyl)-2-methyl-2-cyclohexenone in CDCl₃.

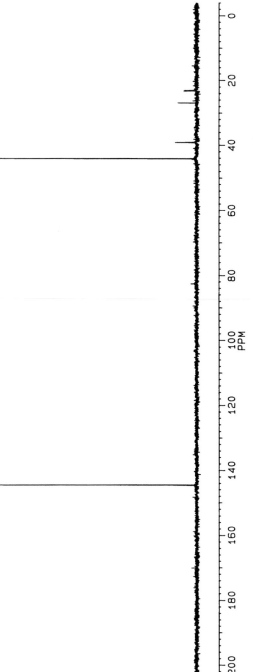

Figure 7.25 The DEPT-90° spectrum of 5-(1-acetoxy-1-methylethyl)-2-methyl-2-cyclohexenone in CDCl₃.

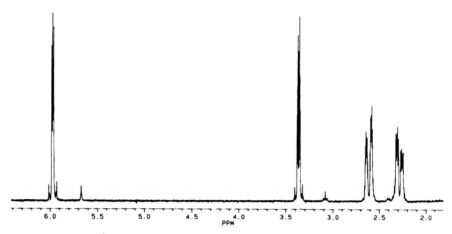

Figure 7.26 The ^{1}H NMR spectrum of 4-cyclohexene-*cis*-1,2-dicarboxylic acid anhydride in $CDCl_3$.

than ^{1}H spectra, and ^{13}C spectra are often better suited for the detection and identification of isomeric or other impurities in a sample; it is easy for small peaks to be concealed underneath a complex second-order multiplet in the ^{1}H NMR spectrum.

The 300-MHz ^{1}H spectrum of 4-cyclohexene-*cis*-1,2-dicarboxylic acid anhydride is shown in Figure 7.26. Owing in part to the presence of two stereocenters, as well as to long-range coupling through the π system of the alkene, the entire ^{1}H spectrum is second order at this field strength, and no information is available from the coupling constants because the spectrum is too complex. Limited assignments to peaks could be made on the basis of chemical shift, but it would be difficult to make any statements regarding purity of our sample based on the ^{1}H NMR spectrum, because an impurity could easily be hidden underneath any of the complex signals.

The ^{13}C spectrum of 4-cyclohexene-*cis*-1,2-dicarboxylic acid anhydride (Figure 7.27) is much less complex. Because of the mirror plane of symmetry in the compound, there are only four different carbon atoms and thus only four lines are seen in the fully decoupled ^{13}C NMR spectrum. This simplicity makes it easy to detect isomeric or other impurities in this sample. These impurities were not as easy to detect in the ^{1}H NMR spectrum. The 1:1:1 triplet centered at 77 ppm is due to the solvent, $CDCl_3$. Although no ^{1}H–^{13}C coupling is observed because of the ^{1}H broadband decoupling, ^{2}H–^{13}C coupling is observed because ^{1}H and ^{2}H resonate at different frequencies.

TWO-DIMENSIONAL NMR SPECTROSCOPY

Two significant developments in NMR spectroscopy are the use of Fourier transform techniques, and the development of two-dimensional (2D) NMR spectroscopy. Two-dimensional spectra are obtained using a sequence of rf pulses that includes a variable delay or delays. A set of FIDs is acquired and

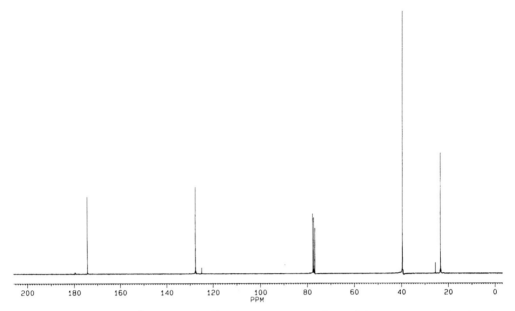

Figure 7.27 Broadband ^{1}H-decoupled ^{13}C NMR spectrum of 4-cyclohexene-*cis*-1,2-dicarboxylic acid anhydride in CDCl$_3$.

stored. The variable delay is incremented by a small amount of time and a new set of FIDs are obtained and stored, and so on. At the end, the resulting matrix of FID data is Fourier transformed twice: once with respect to the acquisition time (as in normal FT-NMR) and second with respect to the time of the variable delay in the pulse sequence. The resulting data represent a surface and are presented as a contour plot of that surface.

The most useful 2D spectra for organic compound identification are called correlation spectra. Correlation spectroscopy (COSY) spectra are presented as a contour plot with routine proton spectra along both of the axes, as shown in the COSY spectrum of ethyl vinyl ether in Figure 7.28. The spectra along the axes are low-digital-resolution spectra and appear to be a bit different from those generated as usual NMR spectra. Note that the 2D spectrum is symmetric about the diagonal that runs from the lower left corner to the upper right corner. Every peak is represented by a peak along the diagonal and, in fact, the diagonal *is* the normal proton spectrum. Where the contour plot indicates a peak other than on the diagonal, the interpretation is that the corresponding peaks on the two axes represent protons coupled to one another.

For example, draw a line down from the signal at 1.2 ppm on the horizontal axis. This line encounters the diagonal at 1.2 ppm and then there is a peak at 3.7 ppm. By drawing a horizontal line over to the spectrum on the vertical axis, we can see that the peaks at 1.2 and 3.7 ppm are coupled to one another; these are the signals from the methyl triplet and methylene quartet of the ethyl group. We can also see that neither of these is coupled to the remainder of the spectrum, which of course is the vinyl group. Each proton in the vinyl group is coupled to each of the others, as we can see in the COSY spectrum. For example, the signal at 4.2 ppm is coupled to both the signal at 4.0 ppm and the one at 6.4 ppm.

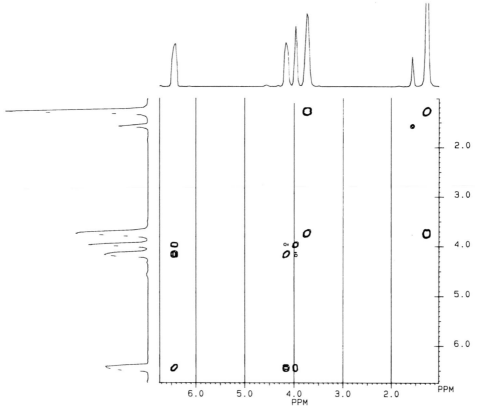

Figure 7.28 The COSY NMR spectrum of ethyl vinyl ether.

It is possible to obtain 2D spectra that correlate the spectra of different nuclei, such as ^{1}H and ^{13}C. The heteronuclear correlation spectrum of ethyl vinyl ether is shown in Figure 7.29. The ^{13}C spectrum is along the horizontal axis and the ^{1}H spectrum is along the vertical axis. The peaks of the 2D spectrum indicate that the corresponding peaks on the axes represent a carbon and a proton (or protons) that are directly bonded. By using this spectrum we can easily verify the ^{13}C and ^{1}H chemical-shift assignments below:

86   14 ppm
152   63

^{13}C chemical shifts

1.25 ppm

4.15 ⟶ H$_A$

H$_B$

H$_C$

3.96   3.73 ppm

6.45

^{1}H chemical shifts

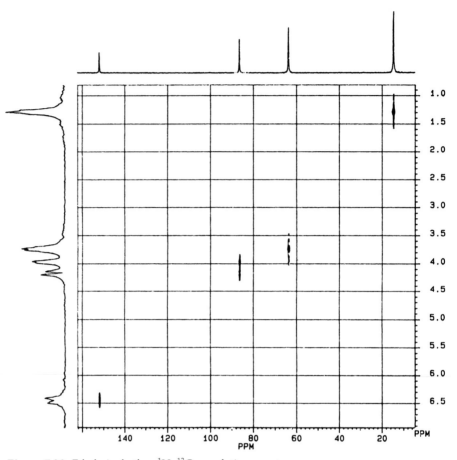

Figure 7.29 Ethyl vinyl ether ^{1}H–^{13}C correlation spectrum.

There are many other powerful 2D NMR spectroscopic techniques that can provide a wealth of information about molecular structure, even in organic molecules as large as proteins and nucleic acids. A few of the good texts that provide further information about these powerful tools are listed in the Bibliography.

NUCLEAR MAGNETIC RESONANCE SAMPLING

It is usually simple to prepare a sample of a small organic molecule (MW 500) for NMR analysis. For ^{1}H NMR spectroscopy, the sample size compatible with CW spectrometers is in the range 30–50 mg dissolved in about 0.5 mL of solvent. Fourier transform spectrometers require only 2–3 mg of sample in the same volume of solvent. Most samples are measured in solution in thin-walled tubes 5 mm in diameter and 18–20 cm long. NMR sample tubes are expensive and delicate; they must be as perfectly straight, and have as perfectly concentric walls, as possible. The sample tube is filled to a depth of about 3 cm. Filling the tube to this depth maximizes sample concentration in the active part of the NMR probe, and thus the strength of the signal. Adding more solvent just wastes sample by dilution. The sample tube is spun about its axis in the instrument to average out small changes in the tube walls and in the magnetic field strength over the sample volume. There must be enough solvent in the tube to ensure that the vortex, or whirlpool, created when the tube is spun, does not extend down into the portion of the tube where the rf coils in the NMR probe are active. Many instruments use a depth gauge that shows exactly where this area is. Glass microcells are now commercially available. An inexpensive microcell technique for CW-NMR spectrometers using ordinary 5-mm NMR tubes has been described.[1]

The most practical NMR solvent is deuterochloroform ($CDCl_3$); it is relatively cheap and dissolves many different compounds. Handle this solvent with care, in the **hood**, because it is **toxic**! Many other deuterated solvents are commercially available, including acetone, methanol, and water. The universally accepted internal reference compound employed in making these measurements is tetramethylsilane, $Si(CH_3)_4$ (TMS). The most convenient source of TMS is commercially available $CDCl_3$, which contains about 1% TMS for use with CW spectrometers (commercially available 0.03% solutions are more appropriate for FT spectrometers).

The most significant problem in sample preparation is the exclusion of small pieces of dust and dirt, because they may contain magnetic material, which will result in poor spectra. Scrupulously clean samples of liquids can often simply be added to the NMR tube, followed by the solvent. Liquids containing visible impurities, and solids, are best prepared by dissolving the sample in about 0.3–0.4 mL of solvent in a small vial. The solution can then be filtered into the NMR tube through a Pasteur pipet plugged with a small piece of cotton, and the pipet can be rinsed with the NMR solvent to achieve the appropriate volume in the NMR tube. Since TMS is volatile (bp = 26.5 °C), the tube should be capped following addition, or the TMS will evaporate. If several people in a laboratory section are going to be obtaining NMR spectra, you will find that the spectrometer will be easier to tune with each new sample if all the sample tubes are filled to exactly the same level.

At this point, all specific instructions are dependent on the NMR spectrometer available to you. In general, the sample is inserted into the magnet and the magnetic field is adjusted very slightly (called shimming or tuning) to obtain a magnetic field that is as homogeneous as possible throughout the sample volume observed by the spectrometer. These adjustments are accomplished by energizing a collection of small electromagnetic coils around the sample, a process that is often done by the spectrometer's computer. Symptoms of a poorly shimmed spectrometer include broad and/or asymmetric peaks. The best place to check for this is on the TMS peak because it is the one peak in the spectrum you *know* should be narrow and symmetric. Once the spectrometer is tuned, the spectrum is obtained and plotted, and the sample is removed from the magnet.

Your NMR sample can be easily recovered by emptying the NMR tube into a small vial, rinsing the tube once or twice with (a nondeuterated) solvent, and then evaporating the solvent in a **hood** under a gentle stream of dry nitrogen.

---

[1] Yu, S. J. *J. Chem. Educ.* **1987,** *64,* 812.

QUESTIONS

Several 60-MHz ^{1}H NMR spectra are given below (Figs. 7.30–7.34) along with the molecular formula of the compound.[2] You should be able to account for at least one acceptable structure and for all of the observed resonances.

7-1. C_4H_8O. Spectrum a:

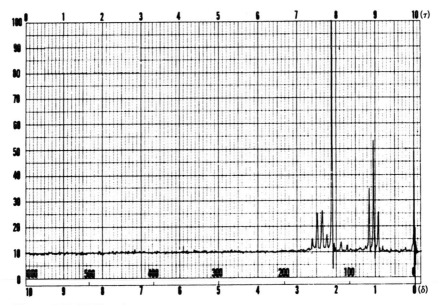

Figure 7.30 NMR unknown spectrum a.

7-2. $C_3H_6O_2$. Spectra b and c. Two compounds with the same empirical formula:

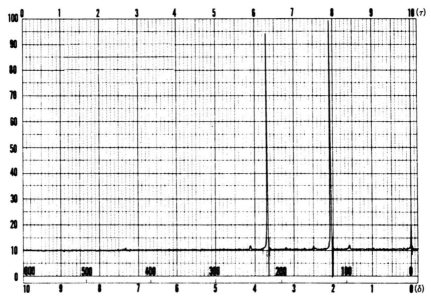

Figure 7.31a NMR unknown spectrum b.

[2]From Pouchert, C. J. *The Aldrich Library of NMR Spectra*, 2nd ed.; Aldrich Chemical Co.: Milwaukee, WI, 1983.

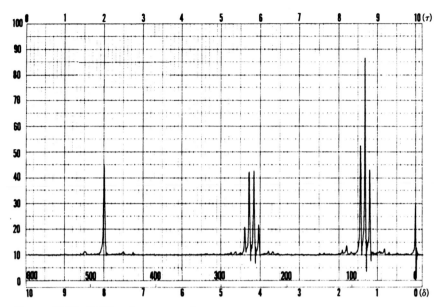

Figure 7.31b NMR unknown spectrum *c*.

7-3. C_4H_8O. Spectrum *d*. Also give some thought to the weak resonances at 0.5 and 1.8 ppm:

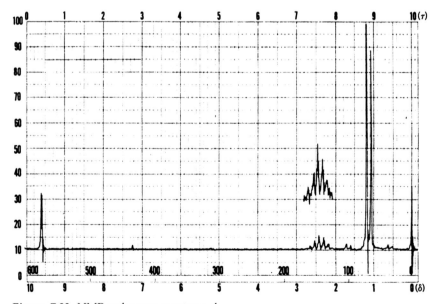

Figure 7.32 NMR unknown spectrum *d*.

7-4. C_7H_7Cl. Spectrum *e*:

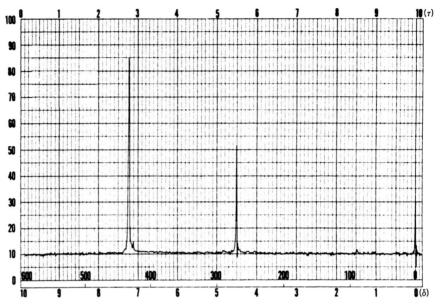

Figure 7.33   NMR unknown spectrum *e*.

7-5. $C_8H_{10}O$. Spectrum *f*:

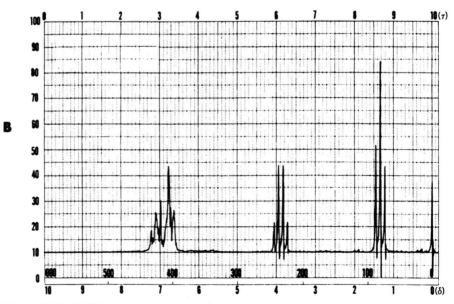

Figure 7.34   NMR unknown spectrum *f*.

BIBLIOGRAPHY

Nuclear magnetic resonance theory and principles of interpretation:

ABRAHAM, R. J.; FISHER, J.; LOFTUS, P. *Introduction to NMR. Spectroscopy*; Wiley: London, 1988.

COOPER, J. W. *Spectroscopic Techniques for Organic Chemists*; Wiley: New York, 1980.

RICHARDS, S. A. *Laboratory Guide to Proton NMR Spectroscopy*; Blackwell Scientific Publications: London, 1988.

SILVERSTEIN; R. M.; WEBSTER, F. X. *Spectrometric Identification of Organic Compounds*, 6th ed.; Wiley: New York, 1997.

SORRELL, T. N. *Interpreting Spectra of Organic Molecules*; University Science Books: Mill Valley, CA, 1988.

STERNHELL, S.; KALMAN, J. R. *Organic Structures from Spectra*; Wiley: London, 1986.

Advanced theory and spectroscopic techniques:

ATTA-UR-RAHMAN *Nuclear Magnetic Resonance. Basic Principles*; Springer-Verlag: New York, 1986.

DEROME, A. E.; *Modern NMR Techniques for Chemistry Research*; Pergamon Press: Oxford, UK, 1987.

DUDDECK, H. DIETRICH, W. *Structure Elucidation by Modern NMR. A Workbook*; Springer-Verlag: New York, 1989.

SANDERS, J. K. M.; HUNTER, B. K. *Modern NMR Spectroscopy. A Guide for Chemists*, 2nd ed.; Oxford University Press: Oxford, UK, 1993.

SANDERS, J. K. M.; CONSTABLE, E. C.; HUNTER, B. K. *Modern NMR Spectroscopy. A Workbook of Chemical Problems*, 2nd ed.; Oxford University Press: Oxford, UK, 1993.

Libraries of NMR spectra:

BHACCA, N. S.; HOLLIS, D. P.; JOHNSON, L. F.; PIER, E. A.; SHOOLERY, J. N. *NMR Spectra Catalog*; Varian Associates: Palo Alto, CA, 1963.

POUCHERT, C. J. *The Aldrich Library of NMR Spectra*, 2nd ed.; Aldrich Chemical Co.: Milwaukee, WI, 1983.

POUCHERT, C. J.; BEHNKE, J. *The Aldrich Library of ^{13}C and ^{1}H FT-NMR Spectra*; Aldrich Chemical Co.: Milwaukee, WI, 1992.

8

INTRODUCTION TO

ULTRAVIOLET–VISIBLE

SPECTROSCOPY

INTRODUCTION TO ABSORPTION SPECTROSCOPY

Theory

In an atom, molecule, or ion, a limited number of electronic energy states are available to the system because of the quantized nature of the energies involved. The absorption of a photon by the system can be interpreted as corresponding to the occupation of a new energy state by an electron. The difference in energy between these two states may be expressed as ΔE:

_____ Upper state (excited electronic state, E_1)

$\Big\uparrow\Big\downarrow \Delta E$

_____ Lower state (ground electronic state, E_0)

where the energy of the photon, E, is related to the frequency of the radiation by the Planck equation,

$$E = h\nu_i$$

where h is Planck's constant, 6.626×10^{-34} J s, and ν_i is the frequency in hertz. In the case above, $\Delta E = E_1 - E_0 = h(\nu_1 - \nu_0) = h\nu_i$.

Thus, when a frequency match between the radiation and an energy gap (ΔE) in the substance occurs, a transition between the two states involved may be induced. The system can either absorb or emit a photon corresponding to ΔE, depending on the state currently occupied (emission would occur if the system relaxed from an upper-level excited state to a lower state). All organic molecules absorb photons with energies corresponding to the visible or ultraviolet regions of the electromagnetic spectrum, but to be absorbed, the incident energy in this frequency range must correspond to an available energy gap between an electronic ground state and an upper-level electronic excited state. The electronic transitions of principal interest to the organic chemist are those that correspond to the excitation of a single electron from the highest occupied molecular orbital (HOMO) to the lowest unoccupied molecular orbital (LUMO). As we will see, this will be the molecule's absorption occurring at the longest wavelength in the electronic absorption spectrum; it is, therefore, the most easily observed.

Electromagnetic radiation can be defined in terms of a frequency ν, which is inversely proportional to a wavelength λ times a velocity c ($\nu = c/\lambda$, where c is the velocity of light in a vacuum, 2.998×10^8 m/s, and $c = \nu\lambda$ is the wave velocity). Thus,

$$\Delta E = h\nu = \frac{hc}{\lambda} = hc\tilde{\nu}$$

where $\tilde{\nu}$ is the wavenumber, defined as the reciprocal of the wavelength $(1/\lambda) \times$ the velocity of light.

Most ultraviolet and visible (UV and vis) spectra are recorded linearly in wavelength, rather than linearly in frequency or in units proportional to frequency (the wavenumber) or in energy values. Wavelength in this spectral region is currently expressed in nanometers (nm, where 1 nm = 10^{-9} m) or angstrom units (Å, where 1 Å = 10^{-10} m). The older literature is full of UV–vis spectra in which wavelength is plotted in millimicrons (mμ), which are also equivalent to 10^{-9} m. For a further discussion of the relationship between frequency, wavelength, wavenumber, and refractive index, see the discussion on infrared spectroscopy.

It is unfortunate that because of instrumentation advantages this region of the spectrum is most often plotted in units that are nonlinear in energy (note the inverse relationship of E to λ). A convenient formula for expressing the relationship of wavelength and energy in useful values is

$$E = 28{,}635/\lambda \text{ kcal/mol} \qquad (\lambda \text{ in nm})$$

or in terms of wavenumbers

$$E = (28.635 \times 10^{-4})\tilde{\nu} \qquad (\tilde{\nu} \text{ in cm}^{-1})$$

The electromagnetic spectrum and the wavelength ranges corresponding to a variety of energy-state transitions are listed in Table 8.1. Infrared, UV–vis, and rf are of particular interest to the organic chemist because the excitation of organic substances by radiation from these regions of the spectrum can yield significant structural information about the molecular system being studied.

TABLE 8.1 *Spectroscopic Wavelength Ranges*

REGION	WAVELENGTH (m)	ENERGY (kJ/mol)	CHANGE EXCITED
Gamma ray	Less than 10^{-10}	$> 10^6$	Nuclear transformation
X-ray	$10^{-8} - 10^{-10}$	$10^4 - 10^6$	Inner shell electron transitions
Ultraviolet (UV)	$4 \times 10^{-7} - 1 \times 10^{-8}$	$10^3 - 10^4$	Valence shell electrons
Visible (vis)	$8 \times 10^{-7} - 4 \times 10^{-7}$	$10^2 - 10^3$	Electronic transitions
Infrared (IR)	$10^{-4} - 2.5 \times 10^{-6}$	$1 - 50$	Bond vibrations
Microwave	$10^{-2} - 10^{-4}$	$10 - 1000$	Molecular rotations
ESR	10^{-2}	10	Electron spin transitions
NMR	$0.5 - 5$	$0.02 - 0.2$	Nuclear spin transitions

The absorption of rf energy by organic molecules immersed in strong magnetic fields involves exceedingly small energy transitions (~ 0.05 cal/mol), which correspond to nuclear spin excitations and result in NMR spectra. When a molecule absorbs microwave radiation, the energy states available for excitation correspond to molecular rotations and involve energies of roughly 1 cal/mol. With relatively simple molecules (in the gas phase) possessing a dipole moment (required for the absorption process) the analysis of the microwave spectrum can yield highly precise measurements of the molecular dimensions (bond lengths and angles). Unfortunately, relatively few organic systems exhibit pure rotational spectra that can be rigorously interpreted.

Absorption of radiation in the infrared region of the spectrum involves the excitation of vibrational energy levels and corresponds to energies in the range of about 1–12 kcal/mol. The excitation of electronic states requires considerably higher energies from a little below 40 to nearly 300 kcal/mol. The corresponding radiation wavelengths would fall across the visible (400–800 nm), the near-UV (200–400 nm), and the far- (or vacuum) UV (100–200 nm) regions. The long-wavelength visible and near-UV regions of the spectrum hold information of particular value to the organic chemist. Here the energies correspond to the excitation of loosely held bonding (π) or lone-pair electrons. The far-UV region, however, involves high-energy transitions associated with the inner-shell and σ-bond electronic energy transitions. This region is difficult to access because atmospheric oxygen begins to absorb UV radiation below 190 nm, which requires working in evacuated or purged instruments (which is why this region is often referred to as the vacuum UV).

UV–VIS SPECTROSCOPY THEORY

As we have seen, the application of electronic absorption spectroscopy in organic chemistry is restricted largely to excitation of ground-state electronic levels in the near-UV and vis regions. When photons of these energies are absorbed, the excited electronic states that result have bond strengths appreciably less than their ground-state values, and the internuclear distances and bond angles will be altered within the region of the molecules where the electronic excitation occurs (see Figure 8.1). It is normally reasonable to assume that nearly all of the molecules are present in the ground vibrational state within the ground electronic state. The upper electronic state also contains a set of vibrational levels and any of these may be open to occupation by the excited electron (see Figure 8.1). Thus, an electronic transition from a particular ground-state level can be to any number of upper-level vibrational states on the excited electronic state.

The shape of an electronic absorption band will be determined to a large extent by the spacing of the vibrational levels and the distribution of band intensity over the vibrational sublevels. In most cases these effects lead to broad absorption bands in the UV–vis region.

The wavelength maximum at which an absorption band occurs in the UV–vis region is generally referred to as the λ_{max} of the sample (where wavelength is determined by the band maximum).

The quantitative relationship of absorbance (the intensity of a band) to concentration is expressed by the Beer–Lambert equation:

$$A = \log \frac{I_0}{I} = \varepsilon c l$$

where A = absorbance, expressed as I_0/I

   I_0 = the intensity of the incident light

   I = the intensity of the light transmitted through the sample

   ε = molar absorbtivity, or the extinction coefficient (a constant characteristic of the specific molecule being observed); values for conjugated dienes typically range from 10,000 to 25,000

   c = concentration (mol/L)

   l = length of sample path (cm)

The calculated extinction coefficient and solvent are usually listed with the wavelength at the band maximum. For example, data for methyl vinyl ketone (3-buten-2-one) would be reported as follows:

λ_{max} 219 nm   (ε = 3600, ethanol)
λ_{max} 324 nm   (ε = 24, ethanol)

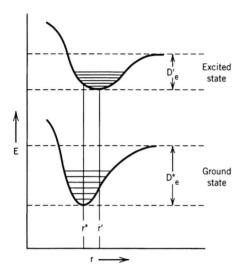

Figure 8.1 Two electronic energy levels in a diatomic molecule.

APPLICATION TO ORGANIC MOLECULES

In organic compounds containing *conjugated* systems of π electrons, a particular *chromophore* present can often be identified by the use of UV–vis spectroscopy. A *chromophore* is, in this case, a group of atoms able to absorb light in the UV–vis region of the spectrum. Since the electronic transitions involved are limited primarily to π-electron (and lone-pair) systems, this type of spectroscopy is less commonly used than the other modern spectroscopic techniques—which, in fact, it predates by several decades. Ultraviolet–visible spectroscopy, however, can play a valuable role in certain situations. For example, if a research problem involves synthesizing a series of derivatives of a complex organic molecule that possesses a strong chromophore, the UV–vis spectrum will be highly sensitive to structural changes involving the arrangement of the π–electron system.

In a conjugated alkene, such as 1,3-butadiene, the long-wavelength photon absorbed corresponds to the energy required for the excitation of a π electron from the HOMO, π_2 to the LUMO, π_3^*. For these alkenes, this transition is represented as $\pi \rightarrow \pi^*$; that is, an electron is promoted from a π (bonding) molecular orbital to a π^* (antibonding) orbital. This type of excitation is depicted below for both ethylene and 1,3-butadiene. Note that as a consequence of extending the chromophore and raising the energy of the highest occupied level in butadiene, the energy gap between the HOMO and LUMO levels of ethylene is larger than that in the conjugated system. Thus, the photon required for excitation of ethylene has a higher energy (higher frequency = shorter wavelength, $\lambda_{max} = 171$ nm) than the photon absorbed by 1,3-butadiene ($\lambda_{max} = 217$ nm):

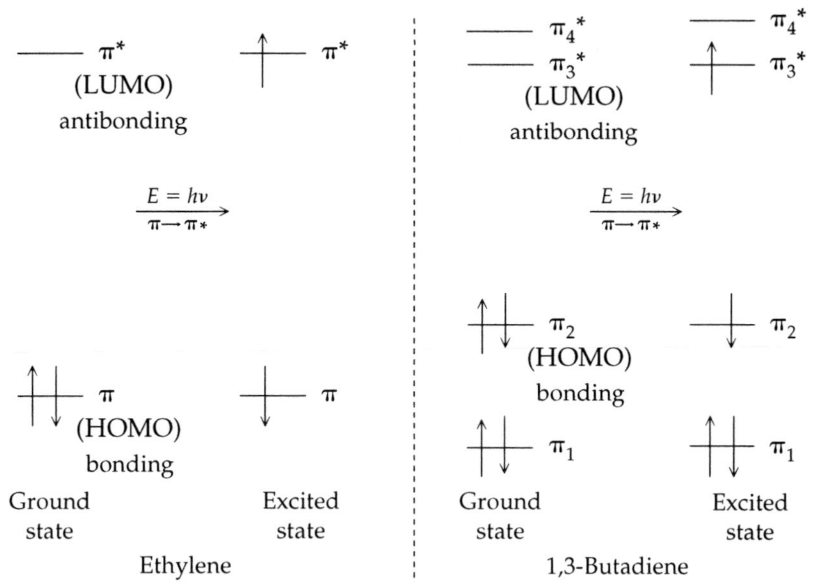

Ethylene | 1,3-Butadiene

If we continue to extend the chromophore, the decrease of the energy gap between the HOMO and LUMO levels also continues. This drop in ΔE is then reflected in a drop in energy of the photon required to excite the $\pi \rightarrow \pi^*$ transition. This effect is illustrated in Table 8.2.

As the extension of the chromophore continues, the λ_{max} of the $\pi \rightarrow \pi^*$ transition will eventually shift into the visible region. At this point the substance exhibits color. Because the absorbed wavelength is coming from the blue end of the visible spectrum, these compounds will appear yellow. The color will deepen and become red as the energy of the photon required for electronic excitation continues to drop.

Compounds that contain a carbonyl chromophore C=O: also absorb radiation in the UV region. A π electron in this unsaturated system undergoes a $\pi \rightarrow \pi^*$ transition. However, unless the carbonyl is part of a more extended chromophore, such as an α,β–unsaturated ketone system, the $\pi \rightarrow \pi^*$ transition requires a fairly high-energy photon for excitation, usually below 190 nm in the far-UV and similar to the energy required for excitation of a carbon–carbon double bond. The edge of the $\pi \rightarrow \pi^*$ absorption band may just barely be observed on instrumentation designed for near-UV studies. This partially observed absorption band is generally referred to as *end absorption*. In the case

TABLE 8.2 *Absorption Maxima of Conjugated Alkene*

NAME	STRUCTURE	λ_{max}(nm)
Ethylene	$CH_2{=}CH_2$	165
1,3-Butadiene	$CH_2{=}CH{-}CH{=}CH_2$	217
1,3,5-Hexatriene	$CH_2{=}CH{-}CH{=}CH{-}CH{=}CH_2$	268
1,3,5,7-Octatetraene	$CH_2{=}CH{-}CH{=}CH{-}CH{=}CH{-}CH{=}CH_2$	290

of carbonyls, however, the heteroatom also loosely holds two pairs of non-bonding electrons that are often termed *lone-pair* electrons. These nonbonding electrons reside in orbitals (n) that are higher in energy than the bonding π orbital, but lower in energy than the antibonding π^* orbital. Thus, while a transition from an n level to a π^* level is formally forbidden, in fact, weak bands are observed at λ_{max} in the near-UV that have their origin in the excitation of a lone-pair electron by an $n \rightarrow \pi^*$ transition. An energy diagram of a typical carbonyl system follows:

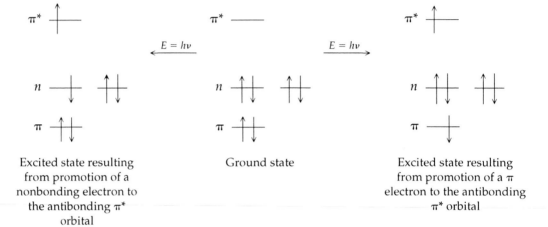

| Excited state resulting from promotion of a nonbonding electron to the antibonding π^* orbital | Ground state | Excited state resulting from promotion of a π electron to the antibonding π^* orbital |

Thus, those substances that contain the carbonyl chromophore absorb radiation of wavelengths that corresponds to both the $n \rightarrow \pi^*$ and the $\pi \rightarrow \pi^*$ transitions. For a simple ketone, such as acetone (CH_3COCH_3), the $\pi \rightarrow \pi^*$ transition is found in the far-UV and the $n \rightarrow \pi^*$ in the near-UV. When the carbonyl becomes part of an extended chromophore, such as in methyl vinyl ketone (3-buten-2-one), the spectra reveal that these two transitions have shifted to longer wavelengths—a bathochromic shift (see Fig. 8.2 for the definition of terms used in UV–vis spectra to indicate the direction of wavelength and intensity shifts):

$$CH_3-\overset{\overset{\displaystyle \cdot\cdot \atop O \cdot\cdot}{\|}}{C}-CH_3 \qquad CH_3-\overset{\overset{\displaystyle \cdot\cdot \atop O \cdot\cdot}{\|}}{C}-CH=CH_2$$

$n \Rightarrow \pi^*$	λ_{max} 270 nm ε_{max} 16	λ_{max} 324 nm ε_{max} 24
$\pi \Rightarrow \pi^*$	λ_{max} 187 nm ε_{max} 900	λ_{max} 219 nm ε_{max} 3600

Saturated systems containing heteroatoms with nonbonded electrons also exhibit weak absorption bands, often as end absorptions, which have their origin in forbidden $n \rightarrow \sigma^*$ transitions. When these heteroatomic groups are attached to chromophores, both the wavelength and the intensity of the absorption can be altered. These are often referred to as *auxochromes* and *auxochromic shifts*.

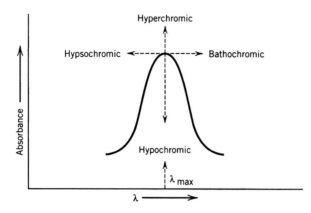

Figure 8.2 Terms describing direction of wavelengths and intensity shifts.

Often, model compounds containing a chromophore of interest are referred to as an aid in the interpretation of the UV–vis spectrum of a new structure. Substantial collections of data have been developed for a wide variety of chromophores as an aid to this type of correlation. A number of empirical correlations, such as the Woodward–Fieser rules, of substituent effects on λ_{max} values are available (see Silverstein et al. and Pasto et al. in the Bibliography). The Woodward–Fieser rules are a set of empirical correlations derived from studies of UV–vis spectral data. Using these rules it is possible to predict with reasonable accuracy the λ_{max} for *new* systems containing various substituents on *known* chromophores. The rules are summarized in Table 8.3.

TABLE 8.3 *Woodward–Fieser Rules for Conjugated Dienes*

FUNCTIONALITY	INCREMENT (nm)
Base value for homoannular diene	253
Base value for heteroannular diene	214
Add:	
For each double bond extending conjugation	+30
For double bond outside of ring (exocyclic)	+5
For alkoxy groups	+6
For S–alkyl groups	+30
For Cl, Br groups	+5
For dialkylamino groups	+60
For parts of rings attached to butadiene fragment	+5

Examples of homoannular and heteroannular dienes are shown below:

A heteroannular diene   A homoannular diene

An example illustrating the use of these rules follows. Calculate the wavelength at which the following steroidal methyl sulfide will absorb:

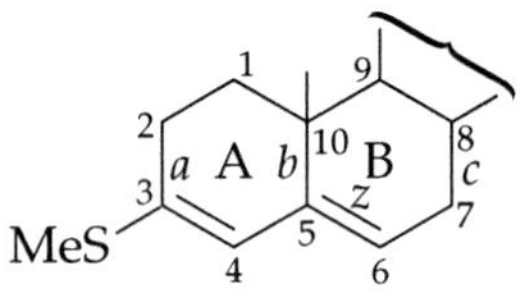

The base value for the diene is 214 nm, because the system is heteroannular (if a homoannular diene were present it would take precedence over the heteroannular diene; see the following example). There are three ring residues (or alkyl substituents) attached to the chromophore. Through hyperconjugation, the π system is slightly extended by this type of substitution. The residues are labeled a, b, and c. Each of these substituents is assumed to add 5 nm to the λ_{max} of the parent heteroannular diene, for a total of 15 nm. The 5,6-double bond in the B ring marked z is exocyclic to the A ring, so empirically we add an additional 5 nm. Finally, for the thiomethyl substituent at the 3 position we add 30 nm. The total is 214 + 15 + 5 + 30 = 264 nm.

Thus we have

Predicted value λ_{max} (calcd) = 264 nm

Observed value λ_{max}(obsd) = 268 nm   (ε = 22,600)

As another example, consider (**I**):

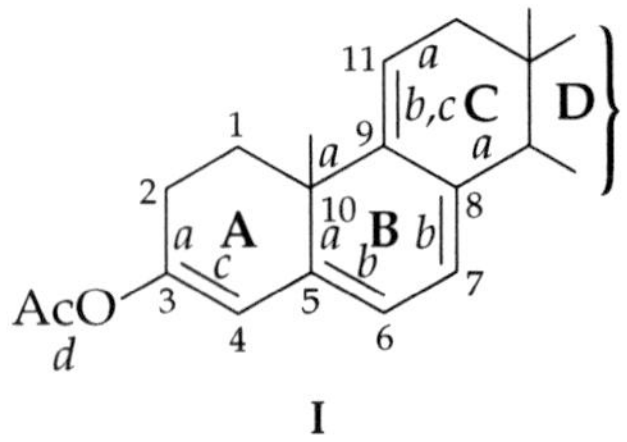

I

Prediction of λ_{max} for a homoannular diene

Parent homoannular diene in ring **B**		253 nm
Increments for		
Double bond extending conjugation	c [2 × 30]	60
Alkyl substituent or ring residue	a [5 × 5]	25
Exocyclic double bond	b [3 × 5]	15
Polar substituents	d [0]	0
	λ_{calcd}	353

Predicted value λ_{max} (calcd) = 353 nm

Observed value λ_{max} (obsd) = 355 nm   (ε = 19,700)

There are additional rules for carbonyl–containing compounds, such as ketones, aldehydes, carboxylic acids, and so on, and for aromatic compounds.

Table 8.4 lists the parameters for conjugated carbonyl systems. Note that in contrast to the conjugated diene compounds, in which we are observing $\pi \to \pi^*$ transitions, the $n \to \pi^*$ transitions of the carbonyl λ_{max} chromophore are often solvent dependent. Thus, solvent effects will have to be considered when predicting λ_{max} values in these systems.

TABLE 8.4 *Conjugated Carbonyl Systems*

α,β-UNSATURATED FUNCTIONALITY			BASE VALUE (nm)
Acyclic or six-membered or higher cyclic ketone			215
Five–membered ring ketone			202
Aldehydes			210
Carboxylic acids and esters			195
			INCREMENT (nm)
Extended conjugation			+30
Homoannular diene			+39
Exocyclic double bond			+5
	Substituent Increment (nm)		
SUBSTITUENT	α	β	δ
Alkyl	+10	+12	+18 (γ and higher)
Hydroxyl	+35	+30	+50
Alkoxy	+35	+30	+31 (γ + 17)
Acetoxy	+6	+6	+6
Dialkylamino		+95	
Chloro	+15	+12	
Bromo	+25	+30	
Alkylthio		+85	
SOLVENT			SOLVENT INCREMENT (nm)
Water			−8
Ethanol			0
Methanol			0
Chloroform			+1
Dioxane			+5
Ether			+7
Hexane			+11
Cyclohexane			+11

An example of λ_{max} (ethanol) calculation for a carbonyl system is presented here:

The base value for the α,β-unsaturated six-membered ring ketone system is 215 nm. Extended conjugation adds an additional 30 nm. The presence of an exocyclic double bond, marked a, extends the λ_{max} another +5 nm. There is a substituent on the β-carbon atom (+12 nm) and on the δ-carbon atom (+18 nm). There is no solvent effect because the spectrum was obtained in ethanol (0 shift). The total is 215 + 30 + 5 + 12 + 18 = 280 nm:

$$\text{Predicted value} \qquad \lambda_{max} \text{ (calcd)} = 280 \text{ nm}$$

$$\text{Observed value} \qquad \lambda_{max} \text{ (obsd)} = 284 \text{ nm}$$

The Woodward–Fieser rules work well for systems with four or fewer double bonds. For more extensively conjugated systems, λ_{max} values are more accurately predicted using the Fieser–Kuhn equation:

$$\text{Wavelength} = 114 + 5M + n(48.0 - 1.7n) - 16.5\,R_{endo} - 10R_{exo}$$

where n = number of conjugated double bonds
M = number of alkyl substituents in the conjugated system
R_{endo} = number of rings with endocyclic double bonds in the system
R_{exo} = number of rings with exocyclic double bonds in the system

Sample calculation: Find the UV λ_{max} of β-carotene:

β-Carotene

In the structure there are 11 conjugated double bonds, $n = 11$. There are six alkyl groups and four ring residues on the conjugated system, $M = 10$. Both rings have an endocyclic double bond, $R_{endo} = 2$. Neither ring has any exocyclic double bonds, therefore $R_{exo} = 0$. Substituting in the equation gives

$$\text{Wavelength} = 114 + 5(10) + 11[48 - 1.7(11)] - 16.5(2) - 10(0)$$

$$= 114 + 50 + 322.3 - 33 - 0 = 453 \text{ nm}$$

$$\text{Predicted value} \qquad \lambda_{max}\text{(calcd)} = 453 \text{ nm}$$

$$\text{Observed value} \qquad \lambda_{max}\text{(obsd)} = 455 \text{ nm}$$

TABLE 8.5 *The Benzoyl Chromophore*

PARENT CHROMOPHORE C_6H_5—CO—R			
FUNCTION			WAVELENGTH (nm)
R = alkyl or ring residue			246
R = H			250
R = OH, O-alkyl			230

	Substituent Increment (nm)		
SUBSTITUENT	o-	m-	p-
Alkyl or ring residue	3	3	10
—OH, —OCH₃, —O-alkyl	7	7	25
—O⁻ (p-sensitive to steric effects)	11	20	78
—Cl	0	0	10
—Br	2	2	15
—NH₂	13	13	58
—NHAc	20	20	45
—NHCH₃			73
—N(CH₃)₂	20	20	85

Note. Spectra obtained in alcohol solvents.

Two examples of this correlation scheme (see Table 8.5) are

1. 6-Methoxytetralone

Predicted λ_{max} is calculated by taking

Parent value, 246 nm + one o-ring residue, 3 + one p-OMe, 25 = 274 nm

Predicted value λ_{max} (calcd) = 274 nm

Observed value λ_{max}(obsd) = 276 nm   (ε = 16,500)

2. 3-Carboethoxy-4-methyl-5-chloro-8-hydroxytetralone

Parent value, 246 nm + one o-ring residue, 3 + one o-OH, 7 + one m-Cl, 0 = 256 nm.

Predicted value λ_{max} (calcd) = 256 nm

Observed value λ_{max} (obsd) = 257 nm $(\varepsilon = 8000)$

For further examples of these types of calculations, see Silverstein et al. and Pasto et al. in the Bibliography.

Table 8.6 lists the λ_{max} values of a number of common organic molecules.

TABLE 8.6 *Absorption Maxima of Several Unsaturated Molecules*

COMPOUND	STRUCTURE	λ_{max}(nm)	ε_{max}
Ethylene	$CH_2{=}CH_2$	171	15,530
1,3-Butadiene	$CH_2{=}CH{-}CH{=}CH_2$	217	21,000
Cyclopentadiene		239	3,400
1-Octene	$CH_3(CH_2)_5CH{=}CH_2$	177	12,600
trans-Stilbene	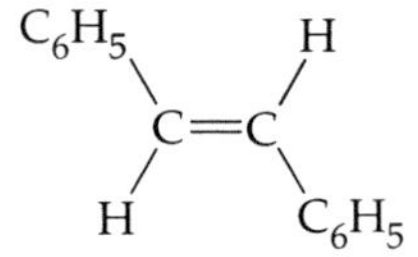	295	27,000
cis-Stilbene	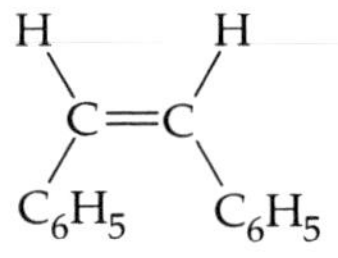	280	13,500
Toluene	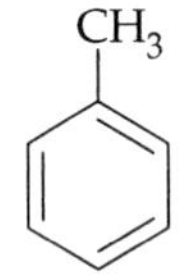	189 208 262	55,000 7,900 260
4-Nitrophenol	 $HO{-}\bigcirc{-}NO_2$	320	9,000
3-Penten-2-one	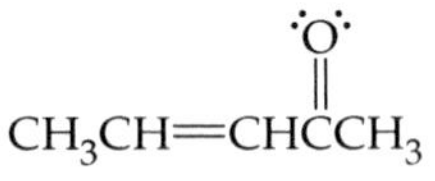 $CH_3CH{=}CHCCH_3$	220 311	13,000 35

In summary, UV–vis spectra can make substantial contributions to understanding the molecular structure of organic substances that possess chromophores:

1. Interpretation of ultraviolet–visible spectra often can be a powerful approach for identifying the molecular structure of that section of a new substance that contains the chromophore.

2. The λ_{max} increases within a series of compounds that contain a common chromophore that is lengthened (increased conjugation) over the series. The intensity of the absorption (ε_{max}) also generally becomes greater as conjugation increases, but can be very sensitive to steric effects.

3. The λ_{max} is sensitive to hyperconjugation by alkyl substituents, conformational changes that restrict π-system overlap, configurational, or geometric isomerization in which π systems are perturbed, and structural changes, such as the isomerization of a double bond from an *exocyclic* to an *endocyclic* position and changes in ring size.

4. In many instances, accurate prediction of the λ_{max} of a new molecular system can be made based on empirical correlations of the parent chromophore giving rise to the absorption.

INSTRUMENTATION

The acquisition of UV–vis absorption spectra for use in the elucidation of organic molecular structure is now carried out with instrumentation that is typically an automatic-recording photoelectric spectrophotometer. The optical components of one of the classic spectrophotometers is given in Figure 8.3. This system is typical of a high-quality double-beam double-monochromator instrument. The instrument consists of a number of components: the radiation source, monochromator, sample compartment, detector, amplifier, and recorder.

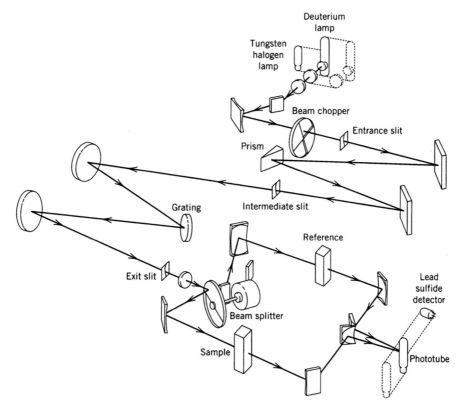

Figure 8.3 Schematic optical diagram of a double beam-in-line spectrophotometer with double monochromation (Cary Model 17D). *(Courtesy of Varian Associates, Inc.)*

The source of radiation

Radiant energy may be generated by either a deuterium discharge lamp or a tungsten–halogen lamp depending on the spectral region to be observed. Deuterium is generally preferred over hydrogen since the intense radiating ball of plasma is slightly larger in the case of deuterium, and therefore source brightness is enhanced by a factor of about 4. Below 360 nm, deuterium gas emits an intense continuum band that covers a major portion of the UV region. With special windows the short wavelength cutoff can be extended down to about 160 nm well out into the vacuum-UV. Emission line spectra limit the long wavelength use of these lamps to about 380 nm. The lamps of choice for the region above 350 nm (the visible) are incandescent filament lamps, because they emit a broad band of radiation from 350 nm on the short wavelength end all the way to about 2.5 μm (the near-IR) on the long wavelength side. Most of the radiation emitted falls outside the visible, peaking at about 1 μm in the near-IR. Nevertheless, tungsten lamps are *the* choice for measurements in the visible region, because they are extremely stable light sources.

Thus, radiation sources must possess two basic characteristics: (1) they must emit a sufficient level of radiant energy over the region to be studied so that the instrument detection system can function, and (2) they must maintain constant power during the measurement period. Source power fluctuations can result in spectral distortion.

The monochromator

As the name implies, a monochromator (making a single color or hue) functions to isolate a single frequency from the source band of radiation. In practice we settle for isolating a small collection of overlapping frequencies surrounding the monochrome radiation we wish to observe. Thus, the monochromator section of the instrument takes all the source radiation in at one end and releases a very narrow set of bands of radiation at the other end. This function is accomplished, as shown in Figure 8.3, by focusing the entering radiation on an entrance slit that forms a narrow image of the source. After passing through the entrance slit, the spreading radiation is collimated by being reflected off a parabolic mirror, and is converted into parallel light rays (just as in a search light). The collimated radiation is then directed to the dispersing agent, which is usually a quartz prism (quartz is transparent to UV, glass is not) or diffraction grating. The dispersing device spreads the different wavelengths of collimated light out in space. After emerging from the prism the dispersed radiation is redirected to either the same or a new collimator mirror and refocused as an image of the source on the exit slit of the monochromator. The exit slit has only a small fraction of the original radiation focused on it, and allows it to pass through in the image of the source. The remaining frequencies lie at different angles on either side of the exit slit. By mechanically turning the prism or grating, and thus changing the angle of the dispersing device with respect to the exit slit, all of the narrowly dispersed bands of radiation can be passed out of the monochromator in sequential fashion.

Instruments that are designed to reduce unwanted radiation to an absolute minimum will place two monochromators in tandem with an intermediate slit connecting the dispersing systems. In the case illustrated in Figure 8.3, the first monochromator uses a prism, while the second uses a grating. The two monochromators, however, must be in perfect synchronization or no light at all will be transmitted.

Sample compartment

After leaving the monochromator the radiation is directed to the sample compartment by a rotating sector mirror, where it is alternately focused on the substance to be examined (which is contained in a cell with quartz windows) and a reference cell (which holds the pure solvent used to dissolve the sample). The system now has two beams, hence the name *double-beam spectrophotometer*. After passing through the sample where the absorption of radiation may occur, the beams are recombined.

The sampling position could be placed either before or after the monochromator. In infrared instruments (such as the PE Model 710B, Fig. 6.3) it was generally found before the monochromator until the introduction of interferometers. In UV systems, the sampling area is most often placed after the monochromator, and for good reason. If the sample were placed before the monochromator, it would be exposed to the entire band of high-energy UV radiation being emitted by the source over the entire sampling period. By positioning the sample after the monochromator, at any one time the sample sees only the very small fraction of the dispersed radiation passed by the exit slit. Thus, sample stability is greatly protected by this arrangement. Remember that near-UV radiation carries photons with energies that approach those of the bond energies of organic molecules.

The detector

The recombined beams are then focused on the detector. Detectors function as transducers because they convert electromagnetic radiation into electrical current. There are a number of radiation-sensitive transducers available as detectors for these instruments. One is the photomultiplier tube. These detectors operate with photocathodes that emit electrons in direct proportion to the number of photons striking the photosensitive surface and possess very large internal amplification. Thus, they operate at low power levels. One particular advantage of the photomultiplier is that you can adjust their sensitivity over a wide range simply by adjusting the supply voltage.

The electronics: The amplifier and recorder

In double-beam instruments; the two signals generated by the sample and reference beams (each referenced against a dark current) in the detector are amplified and the ratio of the sample signal to the reference signal is plotted on a recorder. The simplest of the absorption spectrometers are the single-beam instruments (see Fig. 8.4). These spectrometers are generally employed for problems involving simple one-component analyses. The photometric accuracy of scanned spectra should not be of paramount importance with these systems. Single-beam spectrometers require extremely stable sources and detectors.

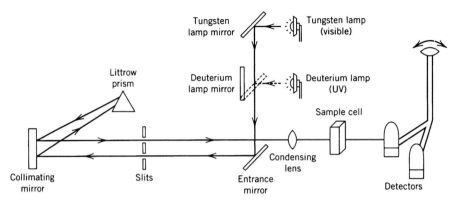

Figure 8.4 UV–visible single-beam spectrometer.

SAMPLE PREPARATION

Ultraviolet spectra are usually obtained on samples in solution using quartz cells. Quartz is used because it is transparent to both UV and visible light. For spectra restricted to the visible region, Pyrex cells are satisfactory (and a good deal less expensive), but because Pyrex absorbs UV radiation, these cells cannot be employed for measurements in this region.

Solution cells usually have a horizontal cross section of 1 cm^2 and require about 3 mL of sample solution. Cells must be absolutely clean, and it is advisable to rinse the cell several times with the solvent used to dissolve the sample. A background spectrum of the solvent-filled cell (*without* a reference sample) can easily be obtained at this time and used as a check against contamination of either the cell or the solvent or both.

Because the intensities of electronic transitions vary over a very wide range, the preparation of samples for UV–vis spectra determination is highly concentration dependent. Intense absorption can result from the high molecular extinction coefficients found in many organic chromophores. The sampling of these materials requires very dilute solutions (on the order of 10^{-6}–10^{-4} M). These solutions can be conveniently obtained by the technique of *serial dilution*. In this method a sample of the material to be analyzed is accurately weighed, dissolved in the chosen solvent, and diluted to volume in a volumetric flask. Sample weights of 4–5 mg in 10-mL volumetric flasks are typical. An aliquot is then taken from this original solution, transferred to a second volumetric flask, and diluted as before. This sequence is repeated until the desired concentration is obtained.

Numerous choices of solvent are available (a list is given in Table 8.7) and most of them are available in "spectral grade." The most commonly used solvents are water, 95% ethanol, methanol, and cyclohexane.

CRITERIA FOR CHOOSING A SOLVENT

⬤ The most important factor is solubility of the sample. UV–vis spectra can be very intense, so even low solubility may be quite acceptable in sample preparation.

TABLE 8.7 *Solvents Used in the Near-UV*

SOLVENT	CUTOFF WAVELENGTH (nm)
Acetonitrile	190
Chloroform	245
(toxic, substitute CH_2Cl_2)	235
Cyclohexane	205
1,4-Dioxane	215
(toxic, substitute EtOEt)	218
95% Ethanol	205
n-Hexane	195
Methanol	205
Isooctane	195
Water	190

Note. Since these solvents have no color, they are transparent in the visible.

- The wavelength cutoff for the solvent may be important if the sample absorbs below about 250 nm.
- Sample–solvent molecular interactions must be considered. An example of these effects would be hydrogen bonding of protic solvents with carbonyl systems. Hydrocarbon chromophores are less influenced by solvent character than are the more polar chromophores.

BIBLIOGRAPHY

American Petroleum Research Institute Project 44, *Selected Ultraviolet Spectral Data*, Vols. I–IV; Thermodynamics Research Center, Texas A&M University: College Station TX, 1945–1977 (1178 compounds).

FEINSTEIN, K. *Guide to Spectroscopic Identification of Organic Compounds*; CRC Press: Boca Raton, FL, 1995.

FIELD, L. D.; STERNHELL, S.; KALMAN, J. R. *Organic Structures from Spectra*, 2nd ed.; Wiley: New York, 1995.

GRASSELLI, J. G.; RITCHEY, W. M. *Atlas of Spectral Data and Physical Constants for Organic Compounds*, 2nd ed.; CRC Press: Cleveland, OH, 1975.

HARWOOD, L. M.; CLARIDGE, T. D. W. *Introduction to Organic Spectroscopy*; Oxford University Press: New York, 1997.

KEMP, W. *Organic Spectroscopy*, 3rd ed.; W. H. Freeman: New York, 1991.

LAMBERT, J. A.; SHURVELL, H. F.; LIGHTNER, D. A.; COOKS, R. G. *Organic Structural Spectroscopy*; Prentice Hall: Englewood Cliffs, NJ, 1998.

LANG, L., Ed. *Absorption Spectra in the Ultraviolet and Visible Region*, Vols. 1–20; Academic Press: New York, 1961–1975; Vols. 21–24; Kreiger: New York, 1977–1984.

MANFRED, H.; MEIER, H.; ZEEH, B.; LINDEN, A. *Spectroscopic Methods in Organic Chemistry*; Thieme Medical Publishing: New York, 1997.

PASTO, D. J.; JOHNSON, C. R.; MILLER, M. J. *Experiments and Techniques in Organic Chemistry*; Prentice Hall: Englewood Cliffs, NJ, 1992.

PAVIA, D. L.; LAMPMAN, G. M.; KRIZ, G. S. *Introduction to Spectroscopy: A Guide for Students of Organic Chemistry*, 2nd ed.; Saunders College: Philadelphia, 1996.

PRETSCH, E.; CLERC, J. T. *Spectra Interpretation of Organic Compounds*; VCH: Wiley: New York, 1997.

SILVERSTEIN, R. M.; WEBSTER, F. X. *Spectrometric Identification of Organic Compounds*, 6th ed.; Wiley: New York, 1997.

Standard Ultraviolet Spectra; Sadtler Research Laboratories: Philadelphia.

UV Atlas of Organic Compounds, Vols. I–IV; Butterworths: London, 1966–1971.

WILLIAMS, D. H.; FLEMING, I. *Spectroscopic Methods in Organic Chemistry*, 4th ed.; WCB/McGraw-Hill: New York, 1989.

9

Introduction to
Mass Spectrometry[*]

TECHNIQUE 13

In comparison with other forms of spectroscopy, such as NMR, IR, or UV–vis, mass spectrometry is unique in terms of how we generate and interpret the spectrum. Instead of monitoring the absorption of electromagnetic radiation in terms of frequency or wavelength, a mass spectrum can be thought of as a snapshot of a rather unconventional organic reaction involving one energetic reactant that decomposes to give a variety of reaction products. By characterizing the composition of this "reaction mixture" we are able to learn the identity of the starting reactant.

The reaction gets started when a molecule in the gas phase is converted to a radical cation by an energetic collision with an electron, as shown below for N_2O:

$$N_2O_{(g)} + e^- \longrightarrow N_2O^{+\cdot} + 2e^-$$

The fact that we have formed a positive ion (cation) with an unpaired electron (radical) becomes important for understanding the decomposition reactions. The process of forming the radical cation yields a collection of energized *molecular ions* that contain a range of internal energies. The molecular ion is produced in a low-pressure environment where it is unable to bump into other molecules. Fragmentation (bond breaking) results in the formation of charged and neutral products. Ideally, intramolecular rearrangements, which could complicate determination of the molecules original structure, do not occur. Depending on characteristics of the molecule and the amount of energy deposited, a variety of fragmentation reactions can take place. For example, the

---

[*]This section has been written by Elizabeth A. Stemmler, Associate Professor of Chemistry, Bowdoin College.

molecular ion ($N_2O^{+\cdot}$) may fragment to give the following products through one- or two-step reactions:

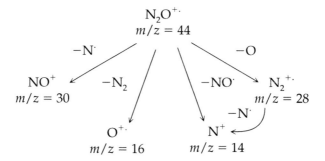

Mass spectrometry derives its name from its ability to distinguish the molecular ion and the different charged reaction products based on the ratio of the ion mass to its charge (m/z ratio). In most cases, the charge, z, is equal to 1 and we can easily tell the difference between the molecular ion ($m/z = 44$) and products, such as N_2^+ (m/z 28). In addition, a mass spectrometer can be used to determine the relative amounts of molecular ions and fragment ions present after the reaction has had a little (very little!) time to proceed. A mass spectrum, typically shown in a bar-graph format, is a display of the relative number of each type of ion plotted as a function of the m/z ratio (see Fig. 9.1). Instead of reporting the actual number of each type of ion, we normalize the data and give the most abundant ion a value of 100%. Note that only charged species are detected by a mass spectrometer. The neutral products are not observed, and their identity must be inferred.

Mass spectrometry is useful to organic chemists because of the information it provides about molecular structure. For example, if the molecular ion is present, that peak can be used to determine the molecular weight (MW) of the neutral molecule. With precise measurement of the m/z ratio (to ± 0.0001, for example), the elemental formula of the molecular ion can be determined. For example, N_2O and CO_2 both have a molecular weight of 44; however, measurement of their exact masses (44.0011 and 43.9898, respectively) can be used

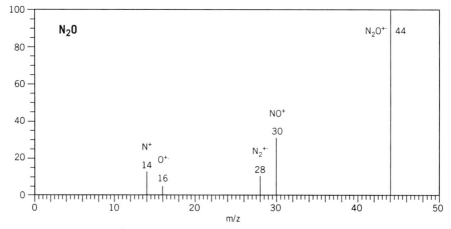

Figure 9.1 Electron ionization mass spectrum of nitrous oxide, N_2O.

to assign their elemental formula. Mass spectrometry was used to originally determine the exact mass of each element (see Table 9.1), and these exact masses, *not* the atomic weights, are used to calculate mass.

Even when precise mass measurements are not available, the products in the mass spectrum may provide enough information to determine the structure of the neutral molecule. Interpretation of a mass spectrum involves working backward from the observed charged fragments to a proposed molecular structure. For example, CO_2 or N_2O (same MW) could be distinguished by an examination of the mass spectrum. The fragment ions at m/z 14 and 30 in Figure 9.1 clearly eliminate CO_2 as a possible structure. There are no combinations of carbon (mass = 12) and oxygen (mass = 16) that could produce these ions. In addition, the mass spectrum allows us to distinguish between the isomers NNO and NON. What would the mass spectrum of NON show? We would expect to see only a peak at m/z 30 (NO^+) and no signal at m/z 28 (N_2^+). The mass spectrum would thus support your chemical intuition that NON is an unlikely and unstable molecular structure.

As you will see below, mass spectral interpretation is not always as straightforward as the case given above. Like the outcome of an organic reaction, a mass spectrum will reflect the outcome of competing sequential and simultaneous reaction pathways. For some molecules, very little fragmentation will take place and only the molecular ion is observed. For other less stable

TABLE 9.1 *Exact Masses and the Atomic Weights for Isotopes of Some Common Elements*

ELEMENT	NUCLIDE	MASS	ATOMIC WEIGHT[a]
Hydrogen	^{1}H	1.0078	1.0079
	D(^{2}H)	2.0141	
Carbon	^{12}C	12.00000 (std)	12.011
	^{13}C	13.0034	
Nitrogen	^{14}N	14.0031	14.0067
	^{15}N	15.0001	
Oxygen	^{16}O	15.9949	15.9994
	^{17}O	16.9991	
	^{18}O	17.9992	
Fluorine	^{19}F	18.9984	18.9984
Silicon	^{28}Si	27.9769	28.0855
	^{29}Si	28.9765	
	^{30}Si	29.9738	
Phosphorus	^{31}P	30.9738	30.9738
Sulfur	^{32}S	31.9721	32.066
	^{33}S	32.9715	
	^{34}S	33.9679	
Chlorine	^{35}Cl	34.9689	35.4527
	^{37}Cl	36.9659	
Bromine	^{79}Br	78.9183	79.904
	^{81}Br	80.9163	
Iodine	^{127}I	126.9045	126.904

[a]Average mass of the naturally occurring isotopes of the element; *not* used for mass calculations in mass spectrometry.

molecules, we may have complete conversion of the molecular ion to products, although, because of the high energy required for their formation, we will rarely see complete fragmentation down to products at the atomic level. The interpretation of a mass spectrum requires developing an understanding of important, characteristic reaction pathways and an appreciation of factors influencing ion stability. In many ways, the interpretation of mass spectra provides a place to apply principles of organic chemistry to a unique kind of chemical reaction.

INSTRUMENTATION

All mass spectrometric instruments contain regions where ionization, mass analysis, and ion detection take place. Mass spectrometry takes place at low pressure; all of the mass spectrometric components are contained in a vacuum system at pressures of 10^{-7} to 10^{-5} torr. Because the instrument must be sealed from the atmosphere to maintain the low pressure, and because samples must be converted to the gas phase prior to ionization, all mass spectrometers have a region devoted to sample introduction. In this region the sample—in the form of a solid, liquid, or gas—is transferred to the low pressure of the mass spectrometer, while preventing the introduction of air. A block diagram of a basic mass spectrometer is shown in Figure 9.2. Ions are generated and fragment in the ion source; the molecular ion and fragments are separated, based upon m/z ratios, in the mass analyzer, and the ion signals are converted by the detector into a signal that may be input to a computer.

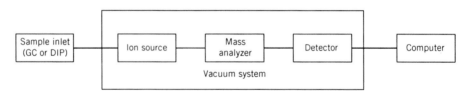

Figure 9.2 Block diagram of components of a mass spectrometer.

Ion source

The ion source is the region where ions are generated. Mass spectrometrists have many ways of creating ions from different types of samples, including biological materials or the surface of a particle. In our discussions, we will focus only on the most common ionization method, electron ionization (EI). In an EI ion source (Fig. 9.3), we send current through a wire, called a filament. As the filament gets hot, electrons are emitted from the surface. The electrons are produced in an electric field, which results in electron acceleration through the ion source region where the sample vapor is found. If you work with an electron energy that is too low (below the ionization potential for the molecule), no ions will be produced. As the electron energy increases, the molecular ion ($M^{+\cdot}$) will appear. With further increases in electron energy, fragment ions are observed. Formation of doubly charged ions (M^{2+}) occurs, but the ion intensities are very low:

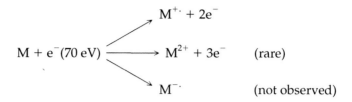

$$M + e^- \, (70\,eV) \longrightarrow \begin{cases} M^{+\cdot} + 2e^- \\ M^{2+} + 3e^- \quad \text{(rare)} \\ M^{-\cdot} \quad\quad\ \text{(not observed)} \end{cases}$$

Mass spectra, by convention, are measured with 70-eV electrons. At this energy, the ion intensity is high and the distribution of products remains relatively constant with small changes in electron energy. Formation of negative molecular ions with 70-eV electrons does not occur.

Another important role for the ion source is directing the ions toward the mass analyzer. The ions are pushed and pulled as they pass through one or more metal plates that have a hole in the center for ion transmission. These plates accelerate the ions and keep them directed at the mass analyzer. Depending on the mass analyzer in use, the ions are accelerated toward the analyzer with high or low energy.

Mass analyzer

Ion formation and fragmentation in the source are followed by mass analysis. Mass analyzers are used to separate ions based on their mass-to-charge ratios. Organic chemists commonly use two types of mass analyzers: magnetic sector

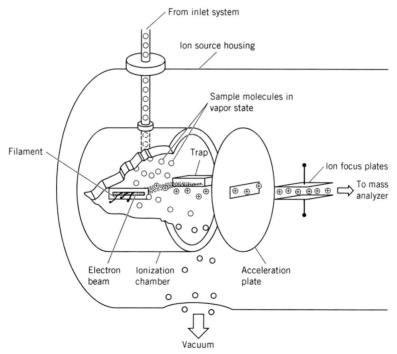

Figure 9.3 Schematic diagram of an electron ionization source. (*From Watson, J. T. Introduction to Mass Spectrometry, 3rd ed.; Lippincott-Raven Publishers; Philadelphia, 1997, p. 140.)*

instruments (low- and high-resolution) and quadrupole instruments. Magnetic sectors separate ions based on dispersion of the ions into beams with different m/z ratios; quadrupoles are mass filtering devices.

In a magnetic sector instrument, ions are accelerated out of the ion source into a magnetic field with high (kilovolt) kinetic energies. The magnet field, applied perpendicular to the path of the ions, exerts a force that causes the ions to follow a curved path through the magnet (Fig. 9.4). The extent to which the path is bent depends on the mass-to-charge ratio (more specifically, the momentum) of the ions. Light ions are bent more than heavier ions. If the path that the ions must travel is fixed, ions that are too light or too heavy will run into the walls of the mass analyzer, where they are neutralized, and will then be pumped away by the vacuum system. Only the ions with the correct radius (correct m/z ratio) will make it to the detector. To measure a complete mass spectrum, the magnetic field strength is varied to bring ions of different m/z ratios to focus on the detector.

High-resolution magnetic sector instruments incorporate an additional energy analyzer prior to mass analysis by the magnetic sector. This more precisely defines the kinetic energies of ions entering the magnetic sector, which improves the mass resolution. High-resolution instruments require more expertise to operate and are less common because of their expense, but they can provide the precise and accurate mass measurements needed to determine elemental composition.

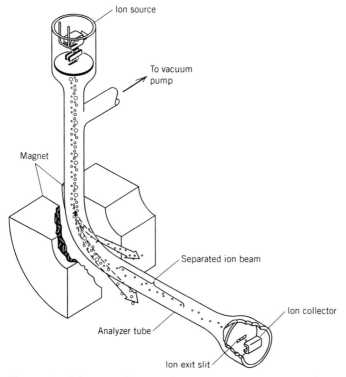

Figure 9.4 Schematic diagram of a magnetic sector mass analyzer. *(From McLafferty, F. W.; Turevek, F. Interpretation of Mass Spectra, 4th ed.; University Science Books: Sausalito, CA, 1993, p. 8.)*

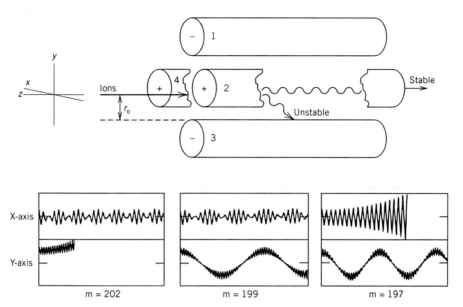

Figure 9.5 Schematic diagram of a quadrupole mass analyzer. X and Y axis trajectories for m/z 202, 199, and 197. *(From Steel, C.; Henchman, M. J. Chem. Educ. **1998**, 75, 1049–1054.)*

A quadrupole is composed of a set of four rods to which (electric) potentials are applied (Fig. 9.5). To allow ions of a particular m/z ratio through the rods, a constant positive potential is applied to two opposing rods (the x-rods), while the remaining two rods experience a constant negative potential (the y-rods). In addition, each set of rods experiences a time varying potential that causes the rod potentials to vary between positive and negative potentials, with the signals 180° out-of-phase. When the x-rods are positive, the y-rods are negative. Mass filtering occurs when a group of ions enters the analyzer. Ions that have m/z ratios that are too low or too high will experience unstable trajectories through the rods and will strike a rod, become neutralized, and be pumped out of the system. Only ions with an appropriate m/z ratio will have a stable trajectory and will make it through the rods to the detector. Figure 9.5 shows the trajectories of three ions with respect to the x- and y-rods. Only the ion with $m/z = 199$ makes it through the quadrupole. To change the m/z ratio of the ions that are transmitted, the magnitude of the constant and time-varying potentials are changed. Mass spectrometers that fit on a laboratory benchtop have a mass range of $m/z = 10$ to 650. With the quadrupole mass analyzer a mass spectrum can be measured rapidly (roughly 1 scan per second), which is important when capillary GC columns are used for sample introduction.

Detector

Ions can be detected directly through the current produced when they strike a plate; however, we usually make this signal larger through the use of electron multiplier detectors.

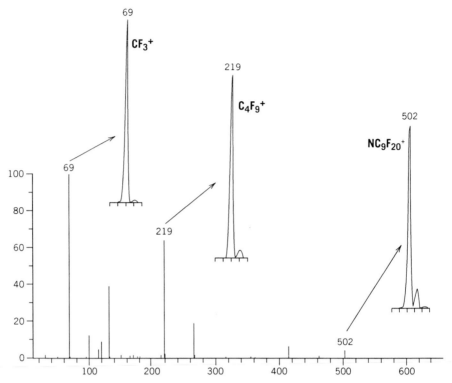

Figure 9.6 Electron ionization mass spectrum of perfluorotributylamine (PFTBA). Inserts show the peak profiles for m/z 69, 219, and 502.

Tuning the mass spectrometer

Before the mass spectrometer can be used to collect mass spectra, the instrument must be tuned and calibrated. The tuning procedure involves setting voltages associated with the ion source, lenses, and detector (to optimize sensitivity), and selecting values for potentials applied to the quadrupole (to set the instrument resolution). These tasks are accomplished while a calibration standard is continuously added to the instrument. A common calibration standard is perfluorotributylamine (PFTBA), $(CF_3CF_2CF_2CF_2)_3N$. Usually ions at m/z 69, 219, and 502 are monitored (Fig. 9.6).

Sample introduction

Samples analyzed by EI mass spectrometry must be converted to gas phase. For pure gases or volatile liquids the samples may be introduced directly through a small orifice that allows an appropriate amount of material into the vacuum chamber. A small amount of a solid sample can be placed in a melting point capillary tube and inserted into the mass spectrometer at the end of a metal rod, called a *direct insertion probe* (DIP). The temperature at the tip of the probe can be varied to promote sublimation of the sample. Another common method of sample introduction is gas chromatography, which is the ideal choice for samples that are impure.

Gas chromatography/mass spectrometry (GC/MS)

While the goal of a synthetic organic reaction is the production of one pure product in high yield, organic reactions often produce a mixture of reaction products. Chromatographic separation of those products is a useful complement to the mass spectral analysis. The components of the mixture elute from the chromatographic column, ideally, as pure peaks. The mass spectrometer, which is scanning rapidly, is then able to collect a few spectra for each eluting peak. Both the chromatographic retention time and the mass spectrum can be used to help identify components of the mixture. Because the compounds are detected with little bias for one type of compound over another, GC/MS has provided organic chemists with a powerful tool to characterize reaction mixtures and assess product purity.

Sensitivity is another distinguishing feature of mass spectrometry. This sensitivity has allowed mass spectrometers to act as detectors for capillary columns, which can separate mixtures containing hundreds of compounds, when less than a nanogram (10^{-9} g) of each compound is injected.

Capillary columns

Most GC/MS instruments use capillary columns for chromatographic separation. Capillary columns are very long (15- to 30-m), open tubes of fused silica that are coated with a thin coating of the stationary phase (Fig. 9.7). A carrier gas, typically helium, is used as the mobile phase. Capillary column diameters are commonly in the range of 0.25 to 0.53 mm, and the coating of the stationary phase is in the range of 0.25 to 1 μm. Thicker coatings are used for the separation of low-boiling compounds.

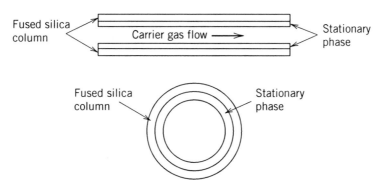

Figure 9.7 Longitudinal and radial cross sections of a capillary column.

Chromatographic resolution increases as a function of the square root of the column length, and the extraordinary length of capillary columns means that most simple mixtures are easily resolved on just a few types of stationary phases. One common nonpolar stationary phase is poly(dimethylsiloxane) (R = CH$_3$), which can be made slightly more polar by the incorporation of phenyl groups (typically 5% phenyl) in place of methyl groups:

$$R-\underset{\underset{R}{|}}{\overset{\overset{R}{|}}{Si}}-O-\left[\underset{\underset{R}{|}}{\overset{\overset{R}{|}}{Si}}-O\right]_n\underset{\underset{R}{|}}{\overset{\overset{R}{|}}{Si}}-R$$

These two stationary phases interact with solutes primarily through dispersion interactions, and compounds elute as a function of boiling point. More polar stationary phases are also available. Because capillary columns are used to separate compounds with a wide range of boiling points, we often make use of a technique called *temperature programming*. This technique allows you to start with a low oven temperature, to optimize the elution of low boiling components, and then increase the oven temperature at a controlled rate, to decrease stationary phase interactions for high-boiling compounds. The increase in temperature decreases retention times and produces narrower peaks for compounds that would otherwise require a long time to elute as a very broad peak.

While samples may be directly injected onto a packed column, the small diameter of the capillary column presents a problem. In addition, it is easy to overload the capillary column with sample (Table 9.2). Two techniques for getting the sample into the column are split and split/splitless injection.

Split injection

In the split injector the sample is injected into the heated injection port and the evaporated sample is mixed with the carrier gas. The sample/carrier gas mixture is then split between the column and a vent, and a fraction of the sample (determined from the column and vent flow) is introduced to the column (Fig. 9.8*a*). This technique is used to introduce concentrated samples.

Split/splitless injection

Splitless injections are used to introduce dilute solutions. The sample is injected into the heated injection port, which is in the "purge off" mode. In this mode, carrier gas flows through the injector directly to the column (Fig. 9.8*b*). This flow rate is very low (0.5–3 mL/min). Of critical importance to splitless injection is the "solvent effect." The oven temperature is maintained below the solvent boiling point, and the vaporized solvent condenses in the column inlet.

TABLE 9.2 *Sample Capacity as a Function of Column Diameter and Stationary Phase Thickness*

COLUMN DIAMETER (mm)	STATIONARY PHASE THICKNESS (μm)	APPROXIMATE CAPACITY (ng/component)
0.25	0.10 (thin)	25
	0.25 (most common)	80
	1.0 (thick)	250
0.53	0.10 (thin)	53
	1.0 (most common)	530
	5.0 (thick)	2600

Source: Alltech Capillary Instruction Manual, Bulletin No. 242; Alltech Associates, Inc.: Deerfield, IL, 1991, p. 9.

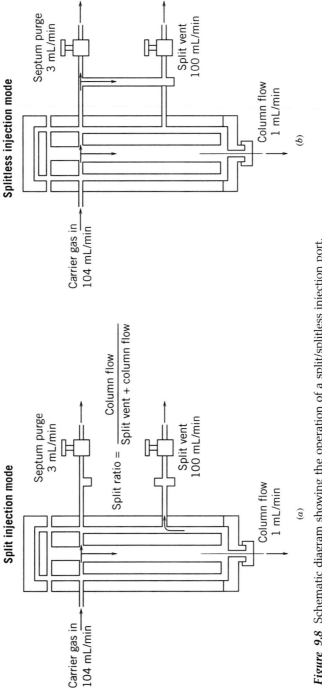

Figure 9.8 Schematic diagram showing the operation of a split/splitless injection port.

This condensed solvent acts like a thick layer of stationary phase and traps sample components. After this concentration period (typically 1 min), the injector is changed to the "purge on" mode. This purge sweeps excess solvent (and other volatile components) out of the injector. Purging too early risks venting volatile components, while purging too late increases interference from the solvent tail.

FEATURES OF THE MASS SPECTRUM

A low-resolution mass spectrum can provide many pieces of information that help an organic chemist determine the structure of a molecule. One of the most useful pieces of information is the compound's nominal molecular weight, MW, as determined by identification of the molecular ion, $M^{+\cdot}$. In addition, by careful examination of the region around the molecular ion for the presence of isotopes, we can learn more about the elemental formula for the molecule. The mass spectrum also reveals information about the molecular structure through the appearance of groups of ions characteristic of certain compound types. With more experience and an understanding of mass spectral fragmentation pathways, a molecular structure can be proposed by gathering all information from the spectrum and determining if a proposed structure is consistent with the observed ions. Here we will provide a limited introduction to this process with an emphasis on identification of the molecular ion. We will present one case study to show how mass spectrometry, coupled with gas chromatography, can be used to characterize the products of a synthetic organic reaction.

Terms

The *molecular ion*, represented by $M^{+\cdot}$, is the intact molecule with one electron missing. This should be the peak in the spectrum with the largest mass, but it is not always observed. All spectra have an ion that we call the *base peak*. This is the most abundant peak in the spectrum (m/z 44 in the spectrum of N_2O or m/z 69 in the spectrum of PFTBA). In the next section you will find that more than one peak may correspond to the $M^{+\cdot}$ or fragment ion when that ion contains elements with different isotopes. We use the term *nominal mass* to describe the mass of the molecule in terms of the most abundant (and, generally, the lightest) isotopes of the element. Relative isotopic abundances for common elements are summarized in Table 9.3.

Isotope peaks

The mass spectrum for $N(C_4F_9)_3$ (PFTBA; MW = 671) is shown in Figure 9.6. The peaks at m/z 69, 219, and 502 are shown above the spectrum as they were measured by the instrument; the spectrum shows their bar-graph representation. These peaks correspond to CF_3^+, $C_4F_9^+$, and $NC_9F_{20}^+$. If you look carefully at the enlarged peaks, you will notice smaller peaks that appear one mass unit above that of the ion. We call these $[A + 1]^+$ peaks. Where do these peaks come from and why does the abundance increase with the mass of the fragment?

If you look at Table 9.3, you will find that fluorine is an isotopically pure element; however, 1.1% of carbon is the ^{13}C isotope. You may recall that it is this low abundance of ^{13}C that you measure with ^{13}C NMR. For nitrogen there

TABLE 9.3 *Relative Isotope Abundances of Common Elements*

ELEMENT	MASS	%	MASS	%	MASS	%
Carbon	12	100	13	1.1		
Hydrogen	1	100	2	0.015		
Nitrogen	14	100	15	0.37		
Oxygen	16	100	17	0.04	18	0.2
Fluorine	19	100				
Silicon	28	100	29	5.10	30	3.4
Phosphorus	31	100				
Sulfur	32	100	33	0.79	34	4.4
Chlorine	35	100			37	32.0
Bromine	79	100			81	97.3
Iodine	127	100				

is a small amount of ^{15}N. When a molecule contains more than one atom of an isotopically impure element, you increase the chance of finding the higher mass isotope in the molecule. For example, the ^{13}C peaks for m/z 69, 219, and 502 of PFTBA (Fig. 9.6) are 1.1, 4.4, and 10.3% of the ^{12}C peaks. The relative intensity increases because there is a higher statistical probability of finding one ^{13}C when the ion has nine vs. one carbon atom. The $[A + 1]^+$ peak intensity from ^{13}C is equal to n times 1.1% the height of the peak A^+, where n is the number of carbon atoms. In addition, we need to add contributions from other $A + 1$ elements, like nitrogen (0.4%). With precise ion intensity measurements, the relative abundance of the $[A + 1]$ peak can be used to determine the number of carbons present in an organic molecule.

Mass spectra get very interesting when chlorine or bromine is present. Both elements exist as mixtures of the A and $A + 2$ isotopes (Table 9.3). The characteristic isotope distribution for bromine, with nearly equal abundances for ^{79}Br and ^{81}Br, is apparent in the spectrum of bromobenzene (Fig. 9.9a). The $M^{+\cdot}$ is observed as a cluster of peaks, with m/z 156 containing the lightest isotopes (^{12}C and ^{79}Br). Note that the fragment at m/z 77 results from loss of Br, and consequently no Br isotope peaks are observed. When more atoms of $A + 2$ elements are present, characteristic peak distributions are produced (Fig. 9.10). For example, an ion that contains two chlorine atoms will show three peaks, A, [A + 2], and [A + 4], with a 100:65:10.6 intensity distribution. Working from Figure 9.10, you can use a pattern recognition approach to determine the number of chlorine or bromine atoms present in an ion. For example, the spectrum in Figure 9.9b shows two ion clusters that suggest the presence of chlorine. A close examination of the distributions indicates that two chlorines are found in the m/z 84 cluster, while the m/z 49 cluster contains one chlorine. When looking at the mass difference between these ions we work with the *nominal* mass (mass of the ion that has the lightest isotopes). The mass difference, defined by the nominal mass of each cluster, is 35, which corresponds to the mass of one ^{35}Cl (Fig. 9.9b).

Recognizing the molecular ion

Compound molecular weight is a valuable piece of information that is not always available from the NMR or IR spectrum. In this section we discuss some

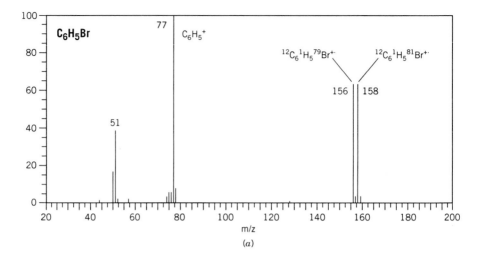

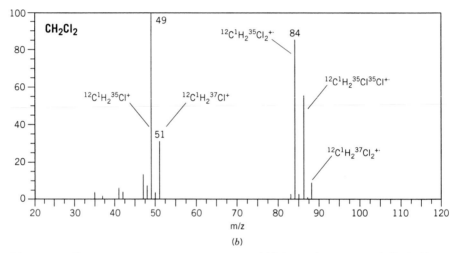

Figure 9.9 Electron ionization mass spectrum of (*a*) bromobenzene and (*b*) dichloro-methane.

things to consider as you examine a mass spectrum and attempt to identify the $M^{+\cdot}$ ion. If you look back at the mass spectra that have appeared above, you will find examples of spectra where the molecular ion is the base peak in the spectrum. In some other cases the molecular ion may be weak or not observed at all! For example, PFTBA fragments extensively and the molecular ion does not appear in the spectrum. The following are some things to consider as you attempt to identify the molecular ion.

The molecular ion should be the highest mass peak in the spectrum. When you have tentatively identified a peak as the molecular ion, you should then determine the masses lost from the molecular ion to give high-mass fragments. For example, in Figure 9.9*b* we found a mass difference of 35 between $M^{+\cdot}$ and the first fragment. A listing of common losses from $M^{+\cdot}$ can be found in Table 9.4. If an observed fragment corresponds to an unreasonable loss, such as M − 12, this strongly suggests that your tentative identification of the molecular ion is incorrect. Remember that it is always possible that no molecular ion is present.

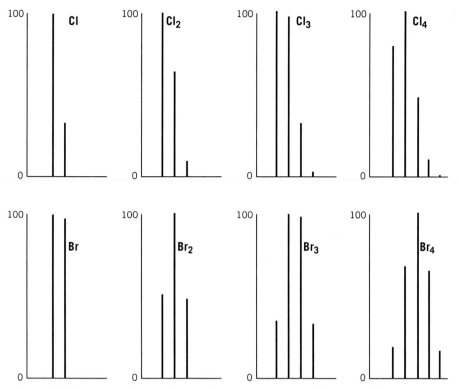

Figure 9.10 Isotope peak distributions for ions containing chlorine and bromine atoms.

Another useful feature to consider is the *nitrogen rule*. For most elements found in organic molecules, the compound molecular weight will be *even* if the compound has an even number of nitrogen atoms (remember, zero is an even number). In contrast, the mass will be *odd* if the compound contains an odd number of nitrogen atoms. If you are sure that your product could not contain nitrogen, then an odd mass ion could not correspond to the molecular ion.

Mass spectrometrists also use "softer" ionization techniques to obtain MW information. These techniques include measuring EI mass spectra at lower

TABLE 9.4 *Some Reasonable[a] Losses From $M^{+\cdot}$*

FRAGMENT[b]	RADICAL LOST	NEUTRAL LOSS
M−1	H	
M−2		H_2
M−15	CH_3	
M−18		H_2O
M−28		CO or C_2H_4
M−29	C_2H_5	
M−31	OCH_3	
M−32		CH_3OH
M−43	C_3H_7	

[a]Unreasonable losses include [M−4] to [M−14]; [M−21] to [M−25].
[b]For a more complete listing see McLafferty and Turecek (1993).

ionization energies and using a higher pressure ionization technique called *chemical ionization.*

Mass spectral interpretation

The following list contains some factors to consider when interpreting mass spectra. To make best use of this summary, the interested reader should consult the text by McLafferty and Turecek, which is considered by many to be the best resource to learn mass spectral interpretation.

1. Using the considerations described above, identify the molecular ion.
2. If possible, determine the elemental composition for $M^{+\cdot}$ and other important peaks using isotopic abundances. In particular, look for isotope peaks from "M + 2" elements like Cl, Br, S, and Si (Table 9.3 and Fig. 9.10). If you are able to establish a molecular formula, calculate "rings + π bonds":

$$\text{For } C_X H_Y N_Z O_N \quad (\text{rings} + \pi \text{ bonds}) = X - \frac{Y}{2} + \frac{Z}{2} + 1$$

Note

> An even-electron ion, with no unpaired electrons, will have a fractional value.

3. Is the molecular weight odd? If so, this indicates an odd number of nitrogen atoms (for organic molecules).
4. Consider the general appearance of the EI mass spectrum: Is it "aliphatic" (lots of fragmentation) or "aromatic" character (minimal fragmentation)?
5. Look for important low-mass ions (Table 9.5).
6. In the region near $M^{+\cdot}$, identify fragments lost from the molecular ion (neutral losses) (Table 9.4). Look for intense high-mass ions that may indicate a characteristic, stable fragment ion.
7. Postulate a structure by assembling the various mass fragments/neutral losses. Do the observed fragment ions make sense in terms of fragment/neutral loss stability considerations? Does the structure make sense

TABLE 9.5 *Some Common Ion Series*

ION SERIES	COMPOUND TYPE
m/z 15, 29, 43, 57, 73	Aliphatic hydrocarbons
m/z 38, 39, 50–52, 63–65, 75–78	Aromatic hydrocarbons (not all peaks in ranges will be observed)
m/z 30, 44, 58	Amines
m/z 31, 45, 59	Alcohols

Note. For a more complete listing see McLafferty and Turecek (1993).

in terms of other information, such as the reaction conditions, NMR or IR spectra?

8. Verify a postulated structure by comparing the spectrum with a reference spectrum. The reference spectrum may be found in the literature or it may be measured by purchasing or synthesizing a standard of the postulated structure.

CASE STUDY: SYNTHESIS OF METHYL BENZOATE[1]

To illustrate how gas chromatography and mass spectrometry can be used to characterize the products of an organic reaction, we consider the synthesis of methyl benzoate using a base-catalyzed esterification of benzoic acid. The reaction proposed for this synthesis involved deprotonation of benzoic acid by n-butyllithium in dry tetrahydrofuran (THF), followed by the addition of methyl iodide, with dimethylformamide (DMF) added to promote the S_N2 displacement of iodide by the benzoate anion:

We isolated the reaction products from the reaction mixture by quenching the reaction with water, adding saturated $NaHCO_3$, and extracting the neutral products with diethyl ether. The ethereal solution containing the reaction products was then analyzed by capillary column GC/MS, and the chromatogram shown in Figure 9.11 was produced. The chromatogram displays total ionization as a function of time. The total ionization is a summation of all the ions detected in one scan of the mass spectrometer (one spectrum) plotted as a point as a function of time. The display is often called the TIC (total ionization chromatogram), and the peak areas should reflect the relative amounts of each compound detected by the mass spectrometer.

The chromatogram shown in Figure 9.11 is not quite what we would hope to see. Instead of detecting a single chromatographic peak for our product, we see three peaks. To determine if we made *any* methyl benzoate, and to determine what other components are present in our mixture, we examine the mass spectrum for each peak. Because we are eager to determine if we made any methyl benzoate, we start by trying to locate a chromatographic peak that has a mass spectrum that corresponds to methyl benzoate. Even if we are not sure what the mass spectrum will look like, we can try to find a spectrum that shows a molecular ion, $M^{+\cdot}$, that corresponds to the molecular weight of methyl benzoate ($C_8H_8O_2$, MW = 136). The mass spectrum for peak 1 (Fig. 9.12a) shows an ion at m/z 136 that appears as the highest mass peak in the spectrum. Let's

---

[1]The synthetic work presented here was carried out by Joshua Pacheco, Bowdoin College Class of 1999.

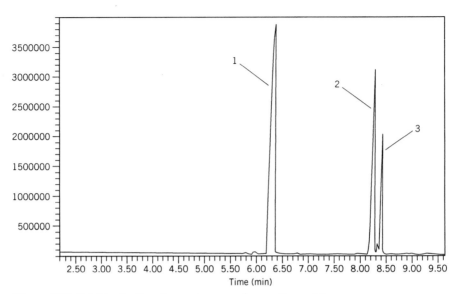

Figure 9.11 Total ionization chromatogram of a reaction mixture.

now take a closer look at the mass spectrum to see if the fragment ions are consistent with the structure of methyl benzoate.

The base peak in the spectrum appears at m/z 105. This intense peak results from a loss of 31 from the molecular ion, which corresponds to loss of OCH_3. This is a predicted loss. Upon ionization, we expect one of the nonbonding electrons on the carbonyl oxygen to be lost, and can consider the charge and unpaired electron to be localized on that oxygen. The unpaired electron initiates a cleavage that results in loss of OCH_3 radical. The ions at m/z 77 and 51, which are characteristic of aromatic rings, may form by cleavage on the other side of the carbonyl group. The ions at m/z 105 and 77 may undergo another fragmentation, but the loss of another radical species is not generally observed from ions of this type, where all electrons are now paired up. Instead, even-electron neutrals, such as CO or C_2H_2, are lost to give fragments m/z 77 and 51, respectively:

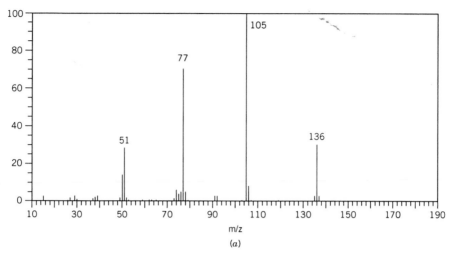

Figure 9.12a Electron ionization mass spectrum of (*a*) peak 1.

Thus, the mass spectrum for peak **1** is consistent with the structure of methyl benzoate. We could further confirm our identification by consulting a library of mass spectra. If we had any pure methyl benzoate around, we could also prepare a standard and use both the GC retention time and the mass spectrum of the standard as means of confirming the compound identification.

We can now move on to some other peaks in the chromatogram. The mass spectra for peaks **2** and **3** (Fig. 9.12*b*, *c*) have many similar features to those of methyl benzoate. Both **2** and **3** show ions at m/z 51, 77, and 105. We can conclude that both compounds have a carbonyl group attached to an aromatic ring. We can next consider identification of the molecular ion. For peak **2**, the highest mass ion is m/z 162. To determine if this is a reasonable assignment for M$^{+\cdot}$, we determine losses from m/z 162. The ions at m/z 133, 120, and 105 could result from losses of 29, 42, and 57, respectively. None of these losses are unreasonable. We next want to consider two important pieces of information. First, if we assume that the MW is 162, we must add 57 to the carbonyl substituted aromatic ring to make our molecule. Addition of a butyl group is a logical choice:

$$\begin{array}{c} \text{105} \quad\Big|\quad \text{57} \\ \text{MW} = 162 \end{array}$$

Next we take note of the peak at m/z 120. The even mass of this ion, resulting from an even mass molecular ion, indicates that it is a special ion! Even-mass fragments generally result from rearrangement reactions, and rearrangements involving hydrogen transfers to carbonyl groups can produce particularly informative product ions. If a butyl group is attached to the

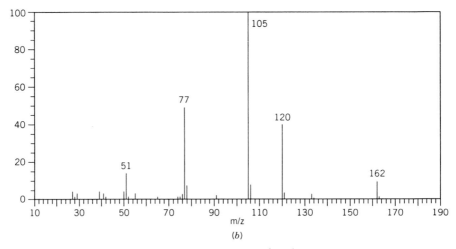

Figure 9.12b Electron ionization mass spectrum of peak 2.

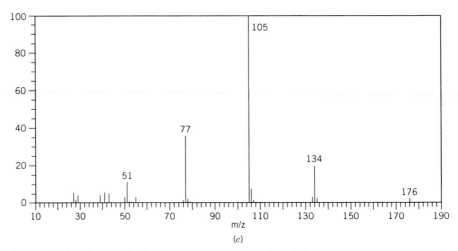

Figure 9.12c Electron ionization mass spectrum of peak 3.

carbonyl, the following fragmentation pathway will occur, which nicely explains the m/z 120 peak:

To determine if this assignment makes sense, we go back to consideration of our reaction. How could this product be generated? If we assume that there is some unreacted n-butyllithium around after the methyl benzoate has been formed, then the following reaction is possible. Thus, we can feel quite confident in our assignment for peak 2:

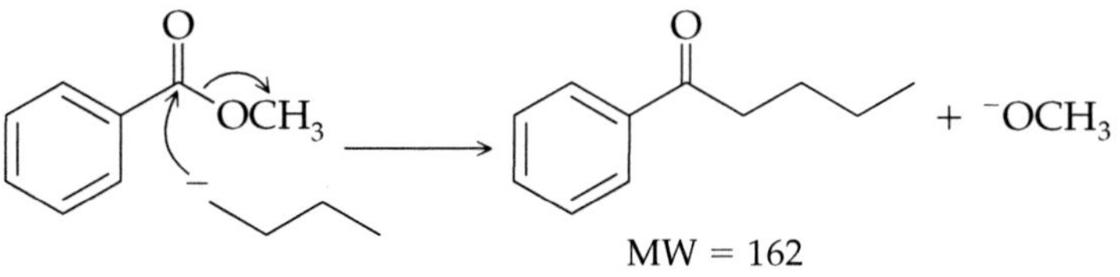

MW = 162

Moving on to peak **3**, we detect an ion at m/z 134. Let's start by assuming that this is our molecular ion. The m/z 105 peak would result from a loss of 29 (C_2H_5) from m/z 134, which is a reasonable loss from $M^{+\cdot}$. This *suggests*, erroneously, that ethyl phenyl ketone is our product:

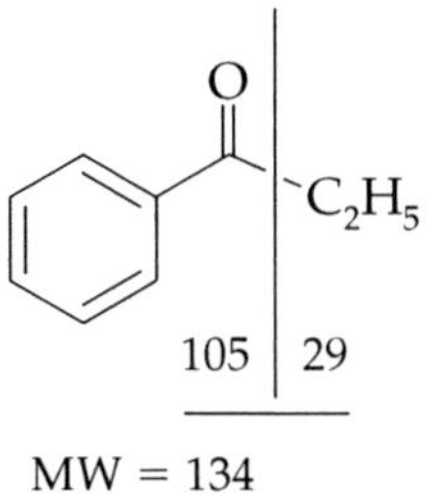

MW = 134

Why is this identification incorrect? If we look back at the chromatogram in Figure 9.11 the chromatographic retention times tell us that something is amiss. Remember that compounds elute from the column in approximate order of increasing boiling point. We would expect that the butyl phenyl ketone would elute *after*, not before, the ethyl phenyl ketone! In addition, you would be hard pressed to propose a mechanism for formation of ethyl phenyl ketone in the context of this reaction. Let's go back and take another look at the mass spectrum. A careful examination shows a small peak at m/z 176. We may have incorrectly identified the molecular ion! If $M^{+\cdot}$ is m/z 176, this gives losses of 42 and 71 to form m/z 134 and 105, respectively. Now it looks like we have a pentyl group attached to the carbonyl, which agrees nicely with the chromatographic retention times. The m/z 134 ion becomes one of our special, even mass ions. What does this ion reveal about the pentyl group? The fact that the ion results from loss of 42, and not 58, clearly indicates that this is not an *n*-pentyl group. What makes the most sense is the branching shown below:

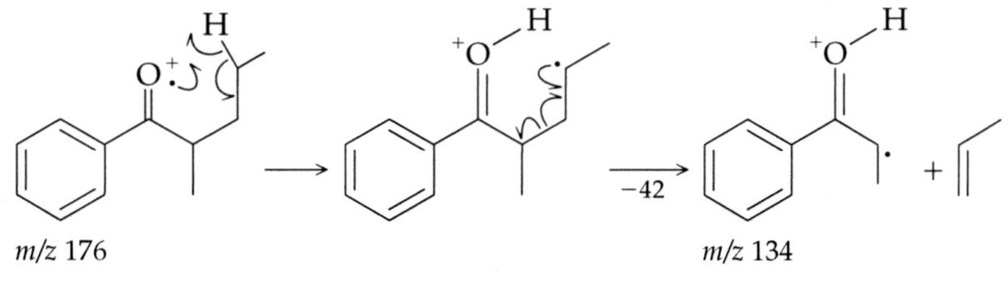

m/z 176 m/z 134

In terms of the chemistry of the reaction, this product also makes sense if deprotonation by *n*-butyllithium occurs α to the carbonyl, followed by reaction with methyl iodide:

MW = 176

BIBLIOGRAPHY

Mass spectral interpretation:

BEYNON, J. H. *The Mass Spectra of Organic Molecules;* Elsevier: Amsterdam, 1968.

BIEMANN, K. *Mass Spectrometry. Organic Chemical Applications;* McGraw-Hill: New York, 1962.

BUDZIKIEWICZ, H.; DJERASSI, C.; WILLIAMS, D. H. *Mass Spectrometry of Organic Compounds;* Holden-Day: San Francisco, 1967.

McLAFFERTY, F. W.; TURECEK, F. *Interpretation of Mass Spectra,* 4th ed.; University Science Books: Sausalito, CA, 1993.

SILVERSTEIN, R. M.; WEBSTER, F. X. *Spectrometric Identification of Organic Compounds,* 6th ed.; Wiley: New York, 1997.

WATSON, J. T. *Introduction to Mass Spectrometry,* 3rd ed.; Lippincott–Raven Publishers: Philadelphia, 1997 (also a good introduction to instrumentation).

Theory and instrumentation:

CHAPMAN, J. R. *Practical Organic Mass Spectrometry;* Wiley: New York, 1985.

HOWE, I.; WILLIAMS, D. H.; BOWEN, R. D. *Mass Spectrometry, Principles and Applications,* 2nd ed.; McGraw-Hill: London, 1981.

KITSON, F. G.; LARSEN, B. S.; McEWEN, C. N. *Gas Chromatography and Mass Spectrometry: A Practical Guide;* Academic Press: San Diego, 1996.

MESSAGE, G. M. *Practical Aspects of Gas Chromatography/Mass Spetrometry;* Wiley: New York, 1984.

Quadrupole Mass Spectrometry and Its Applications; Dawson, P. H., Ed., Elsevier: New York, 1976.

ROBOZ, J. *Introduction to Mass Spectrometry, Instrumentation and Techniques;* Wiley-Interscience: New York, 1968.

Libraries:

McLAFFERTY, F. W. *Registry of Mass Spectral Data,* 5th ed.; Wiley: New York, 1989.

McLAFFERTY, F. W.; STAUFFER, D. B. *Wiley/NBS Registry of Mass Spectral Data;* Wiley: New York, 1989.

NIST/EPA/MSDC Mass Spectral Data Base; National Institute of Standards and Technology. You may also access some mass spectral data through the Internet: *WebBook.nist.gov*

10

Qualitative Identification of Organic Compounds

ORGANIC QUALITATIVE ANALYSIS

One of the exciting challenges that a chemist faces on a regular basis is identifying organic compounds. This challenge is an excellent way for a student to be initiated into the arena of chemical research. Millions of organic compounds are recorded in the chemical literature. At first glance it may seem a bewildering task to attempt to identify one certain compound from this vast array, but most of these substances can be grouped, generally by functional groups, into a comparatively small number of classes. In addition, chemists have an enormous database of chemical and spectroscopic information, which has been correlated and organized over the years, at their disposal. Determination of the physical properties of a molecule, the functional groups present, and the reactions the molecule undergoes has allowed the chemist to establish a systematic, logical identification scheme.

Forensic chemistry, the detection of species causing environmental pollution, the development of new pharmaceuticals, progress in industrial research, and development of polymers all depend to a large extent on the ability of the chemist to isolate, purify, and identify specific chemicals. The objective of organic qualitative analysis is to place a given compound, through screening tests, into one of a number of specific classes, which in turn greatly simplifies the *identification* of the compound. This screening is usually done by using a series of preliminary observations and chemical tests, in conjunction with the instrumental data that developments in spectroscopy have made available to the analyst. The advent of infrared (IR) and nuclear magnetic resonance (NMR) spectroscopy and mass spectrometry (MS) have had a profound effect on the approach taken to identify a specific organic compound. Ultraviolet (UV) spectra may also be utilized to advantage with certain classes of materials.

The systematic approach taken in this text for the identification of an unknown organic compound is as follows:

1. Preliminary tests are performed to determine the physical nature of the compound.

2. The solubility characteristics of the unknown species are determined. This identification can often lead to valuable information related to the structural composition of an unknown organic compound.

3. Chemical tests, mainly to assist in identifying elements other than C, H, or O, may also be performed.

4. Classification tests are carried out to detect *common functional groups* present in the molecule. Most of these tests may be done using a few drops of a liquid or a few milligrams of a solid. An added benefit to the student, especially in relation to the chemical detection of functional groups, is that a vast amount of chemistry can be *observed* and *learned* in performing these tests. The successful application of these tests requires that you develop the ability to think in a logical manner and to interpret the significance of each result based on your observation. Later, as the *spectroscopic techniques* are introduced, *the number of chemical tests performed are usually curtailed*.

5. The spectroscopic method of analysis is utilized. As your knowledge of chemistry develops, you will appreciate more and more the revolution that has taken place in chemical analysis over the past 25–30 years and the powerful tools now at your disposal for the identification of organic compounds. In the introductory laboratory, the techniques of IR, NMR, and UV-vis spectroscopy, and mass spectrometry are generally explored.

It is important to realize that *negative* findings are often as important as *positive* results in identifying a given compound. Cultivate the habit of following a *systematic pathway or sequence* so that no clue or bit of information is lost or overlooked along the way. It is important also to develop the *attitude* and *habit* of planning ahead. Outline a logical plan of attack, depending on the nature of the unknown, and follow it. As you gain more experience in this type of investigative endeavor, the planning stage will become easier.

At the *initial* phase of your training, the unknowns to be identified will be relatively pure materials and will all be known compounds. The properties of these materials are recorded in the literature, and/or in the tables on the website; see ⬛ Appendix. Later, perhaps, mixtures of compounds or samples of commercial products will be assigned for separation, analysis, and identification of the component compounds.

Record all observations and results of the tests in your laboratory notebook. Review these data as you execute the sequential phases of your plan. This method serves to keep you on the path to success.

A large number of texts have been published on organic qualitative analysis. Several references are cited here.

BIBLIOGRAPHY

CHERONIS, N. D.; MA, T. S. *Organic Functional Group Analysis by Micro and Semimicro Methods*; Interscience: New York, 1964.

CHERONIS, N. D.; ENTRIKIN, J. B.; HODNETT, E. M. *Semimicro Qualitative Organic Analysis*, 3rd ed.; Interscience: New York, 1965.

FEIGL, F.; ANGER, V. *Spot Tests in Organic Analysis*, 7th ed.; Elsevier: New York, 1966.

KAMM, O. *Qualitative Organic Analysis*, 2nd ed.; Wiley: New York, 1932.

PASTO, D. J.; JOHNSON, C. R.; MILLER, M. J. *Experiments and Techniques in Organic Chemistry*; Prentice Hall: Englewood Cliffs, NJ, 1992.

SCHNEIDER, F. L. In *Qualitative Organic Microanalysis*, Vol. II of *Monographien aus dem Gebiete der qualitativen Mikroanalyse*; Benedetti-Pichler, A. A., Ed.; Springer-Verlag: Vienna, 1964.

SHRINER, R. L.; HERMANN, C. K. F.; MORRILL, T. C.; CURTIN, D. Y.; FUSON, R. C. *The Systematic Identification of Organic Compounds*, 7th ed.; Wiley: New York, 1998.

VOGEL, A. I. *Qualitative Organic Analysis*, Part 2 of *Elementary Practical Organic Analysis*; Wiley: New York, 1966.

PRELIMINARY TESTS

Preliminary tests help you select a route to follow to ultimately identify the unknown material at hand. These tests frequently consume material, so, given the amounts of material generally available at the micro- or semimicroscale level, judicious selection of the tests to perform must be made (in some tests, the material analyzed may be recovered). Each preliminary test that can be conducted with *little expenditure of time and material* can offer valuable clues as to which class a given compound belongs.

Nonchemical tests

PHYSICAL STATE

If the material is a *solid*, a few milligrams of the sample may be viewed under a magnifying glass or microscope, which may give some indication as to the homogeneity of the material. Crystalline shape is often an aid in classifying the compound.

Determine the melting point, using a small amount of the solid material. A narrow melting point range (1–2 °C) is a good indication that the material is quite pure. If a broad range is observed, the compound must be recrystallized from a suitable solvent before proceeding. If the material undergoes decomposition on heating, try an evacuated (sealed-tube) melting point. If any evidence indicates that sublimation is occurring, an evacuated melting point should be run. Furthermore, this result indicates that sublimation might be used to purify the compound, if necessary.

If the material is a *liquid*, the boiling point is determined by the ultramicro method. If sufficient material is on hand and the boiling point reveals that the material is relatively pure (narrow boiling point range), the *density* and the *refractive index* can provide valuable information for identification purposes.

COLOR

Since the majority of organic compounds are colorless, examination of the color can occasionally provide a clue as to the nature of the sample. Use caution, however, since tiny amounts of some impurities can color a substance. Aniline is a classic example. When freshly distilled it is colorless, but on standing a small fraction oxidizes and turns the entire sample a reddish-brown color.

Colored organic compounds contain a *chromophore*, usually extended conjugation in the molecule. For example, 1,2-dibenzoylethylene is yellow; 5-nitrosalicylic acid is light yellow; tetraphenylcyclopentadienone is purple.

Can you identify the chromophores that cause these compounds to be colored? Note that a colorless liquid or white solid would not contain these units. Thus, compounds containing these groupings would be excluded from consideration as possible candidates in identification of a given substance.

ODOR

Detection of a compound's odor can occasionally be of assistance, since the vast majority of organic compounds have no definitive odor. You should become familiar with the odors of the common compounds or classes. For example, aliphatic amines have a fishy smell; benzaldehyde (like nitrobenzene and benzonitrile) has an almond odor; esters have fruity odors. Common solvents, such as acetone, diethyl ether, and toluene, all have distinctive odors. Butyric and caproic acids have rancid odors. Low molecular weight mercaptans (—SH) have an intense smell of rotten eggs. In many cases, extremely small quantities of certain relatively high molecular weight compounds can be detected by their odor. For example, a C_{16} unsaturated alcohol released by the female silkworm moth elicits a response from male moths of the same species at concentrations of 100 molecules/cm^3. Odors are an important facet of chemical communication between plants and animals and often result in a spectacular behavioral response.

Odor detection in humans involves your olfactory capabilities and thus can be a helpful lead, *but very rarely can this property be used to strictly classify or identify a substance.* As mentioned above, contamination by a small amount of an odorous substance is always a possibility.

Caution

You should be very cautious when detecting odors. Any odor of significance can be detected several inches from the nose. Do not place the container closer than this to your eyes, nose, or mouth. Open the container of the sample and gently waft the vapors toward you.

Ignition test

Caution

Make sure you are wearing safety glasses.

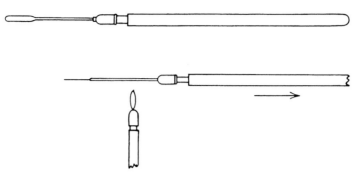

Figure 10.1 Heating on the microspatula. *(Courtesy of Springer-Verlag, Vienna.)*

Valuable information can be obtained by carefully noting the manner in which a given compound burns. The ignition test[1] is carried out by placing 1–2 mg of the sample on a spatula, followed by heating and ignition with a microburner flame. Do not hold the sample directly in the flame; heat the spatula about 1 cm from the flat end and move the sample slowly into the flame (see Fig. 10.1).

Important observations to be made concerning the ignition test are summarized in Table 10.1.

TABLE 10.1 *Ignition Test Observations*

TYPE OF COMPOUND	EXAMPLE	OBSERVATION
Aromatic compounds, unsaturated, or higher aliphatic compounds	Toluene	Yellow, sooty flame
Lower aliphatic compounds	Hexane	Yellow, almost nonsmoky flame
Compounds containing oxygen	Ethanol	Clear bluish flame
Polyhalogen compounds	Chloroform	Generally do not ignite until burner flame applied directly to the substance
Sugars and proteins	Sucrose	Characteristic odor
Acid salts or organometallic compounds	Ferrocene	Residue

Source: Cheronis, N. D.; Entrikin, J. B. *Semimicro Qualitative Analysis*; Interscience: New York, 1947, p. 85.

As the sample is heated, you should make the following observations:

1. Any melting or evidence of sublimation: This observation gives an approximate idea of the melting point by the temperature necessary to cause melting.

2. Color of the flame as the substance begins to burn (see Table 10.1).

3. Nature of the combustion (flash, quiet, or an explosion). Rapid, almost instantaneous combustion indicates high hydrogen content. Explosion indicates

---

[1]For an extensive discussion on examination of ignition residues see Feigl, F.; Anger, V. *Spot Tests in Organic Analysis*, 7th ed.; Elsevier: New York, 1966, p. 51.

the presence of nitrogen- or nitrogen–oxygen-containing groups; for example, nitro groups.

4. Nature of the residue, if present, after ignition.

    a. If a black residue remains and disappears on further heating at higher temperature the residue is carbon.

    b. If the residue undergoes swelling during formation, the presence of a carbohydrate or similar compound is indicated.

    c. If the residue is black initially but still remains after heating, an oxide of a heavy metal is indicated.

    d. If the residue is white, the presence of an alkali or alkaline earth carbonate or SiO_2 from a silane or silicone is indicated.

SEPARATION OF IMPURITIES

If the preliminary tests outlined above indicate that the unknown in question contains impurities, it may be necessary to carry out one of several purification steps. These techniques are discussed in earlier chapters and are summarized below for correlation purposes:

1. For a liquid, distillation is generally used (see Techniques 2 and 3).

2. For a solid, recrystallization is generally used (see Technique 5).

3. Extraction is used if the impurity is insoluble in a solvent in which the compound itself is soluble (see Technique 4).

4. Sublimation is a very efficient technique if the compound sublimes (see Technique 9).

5. Chromatography (gas, column, and thin-layer) is often used (see Techniques 1 and 6A).

These techniques may be applied to the separation of mixtures as well.

DETECTION OF ELEMENTS OTHER THAN CARBON, HYDROGEN, OR OXYGEN

Other than C, H, and O, the elements that are most often present in organic compounds are nitrogen, sulfur, and the halogens (F, Cl, Br, or I). To detect the presence of these elements, the organic compound is generally fused with metallic sodium. This reaction converts these heteroatoms to the water-soluble inorganic compounds, NaCN, Na_2S, and NaX. Inorganic qualitative analysis tests enable the investigator to determine the presence of the corresponding anions:

$$\text{Organic compound containing} \begin{Bmatrix} C \\ H \\ O \\ N \\ S \\ X \end{Bmatrix} \xrightarrow[\Delta]{Na} \begin{Bmatrix} NaCN \\ Na_2S \\ NaX \end{Bmatrix}$$

Sodium fusion[2]

Note

The fusion reaction is carried out in the **hood.** Make sure you are wearing safety glasses. All reagents must be of analytical grade, and deionized water must be used.

Caution

Sodium metal can cause serious burns, and it reacts violently with water.

In a small (10 × 75-mm) test tube (soft glass preferred), supported in a transite board (see Fig. 10.2), place about 25–30 mg of clean sodium metal (about one-half the size of a pea).

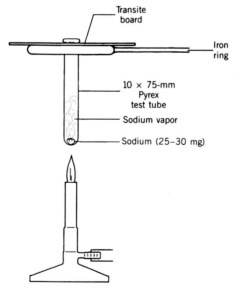

Transite board

Iron ring

10 × 75-mm Pyrex test tube

Sodium vapor

Sodium (25–30 mg)

Figure 10.2 Apparatus for sodium fusion.

[2]See Campbell, K. N.; Campbell, B. K. *J. Chem. Educ.* **1950,** *27,* 261 for a discussion of the procedure.

Caution

> Use forceps to make this transfer; never touch sodium metal with your fingers.

Heat the tube with a flame until the sodium melts and sodium vapor is observed rising in the tube (see Fig. 10.2).

Mix a small sample of your unknown compound (1–2 drops of a liquid; 6–10 mg if a solid) with about 15–25 mg of *powdered* sucrose.[3] Gentle mixing of solids may be done on filter paper or glassine weighing paper; liquids can be mixed on a watch glass. Add this mixture to the tube, being careful not to get any material on the sides of the test tube.

Note

> The addition of sucrose to the sample helps reduce various nitrogen or sulfur compounds. It also absorbs volatile materials so that they may undergo the desired reaction before significant vaporization can occur.

Now heat the tube gently to initiate the reaction with sodium. Remove the flame until the reaction subsides, and then heat to redness for 1–2 min. Allow the tube and contents to cool to room temperature. Then, *and only then*, **cautiously** add several drops of methanol (using a Pasteur pipet) to decompose any unreacted metallic sodium. Gently warm the mixture to drive off the excess methanol.

Reheat the tube to a bright red. While the tube is still red hot, lift the transite board and test tube from the iron ring and place the tube in a small beaker (30 mL) containing about 15 mL of deionized water (the transite board acts as a cover on the beaker).

Caution

> The soft-glass tube usually cracks and breaks during this operation.

Break up the tube with a glass rod, heat the solution to boiling and filter it by gravity into a clean 50-mL Erlenmeyer flask. Wash the filter paper with an additional 2.0 mL of distilled water and combine this wash with the original filtrate.

---

[3]Ordinary confectioner's sugar purchased at the supermarket can be used.

Note

If a Pyrex test tube is used, after the unreacted sodium metal is completely destroyed by adding methanol, add 2 mL of deionized water directly to the tube and contents. Place a glass stirring rod in the tube and heat the solution to boiling with stirring and then filter as described above. Dilute the filtrate with deionized water to about 5 mL.

USING THE FUSION SOLUTION

The clear, colorless fusion solution is used to test for the presence of CN^- (nitrogen), S^{2-} (sulfur), and X^- (halogens, except F^-) as described in the following sections.

Sulfur

1. Place 2–3 drops (Pasteur pipet) of the fusion solution on a white spot plate, followed by 2 drops of water. Now add 1 drop of dilute (2%) aqueous sodium nitroprusside solution. The formation of a deep blue-violet color is a positive test for sulfur:

$$Na_2S + Na_2Fe(CN)_5NO \rightarrow Na_4[Fe(CN)_5NOS]$$

Sodium nitroprusside   Blue-violet complex

2. Place 3–4 drops (Pasteur pipet) of the fusion solution on a white spot plate followed by 1–2 drops of acetic acid. Now add 1 drop of 1% lead(II) acetate solution. The formation of a black precipitate (lead sulfide) indicates the presence of sulfur.

Nitrogen[4]

REAGENTS

    1. A 1.5% solution of *p*-nitrobenzaldehyde in 2-methoxyethanol

    2. A 1.7% solution of *o*-dinitrobenzene in 2-methoxyethanol

    3. A 2.0% solution of NaOH in distilled water

Note

All reagent drops are dispensed using Pasteur pipets.

---

[4]Adapted from Guilbault, G. G.; Kramer, D. N. *Anal. Chem.* **1966**, *39*, 834. *Idem. J. Org. Chem.* **1966**, *31*, 1103. See also Shriner, R. L.; Fuson, R. C.; Morrill, T. C. *The Systematic Identification of Organic Compounds*, 6th ed.; Wiley: New York, 1980, p. 80.

On a white spot plate, place together: 5 drops of reagent **1,** 5 drops of reagent **2,** and 2 drops of reagent **3.** Stir this mixture gently with a glass rod.

Now add 1 drop of the fusion solution. The formation of a deep-purple color is a positive test for the presence of CN⁻ ion; a yellow or tan coloration is negative. If a positive result is obtained, nitrogen is present in the sample.

The test is valid in the presence of halogens (NaX) or sulfur (Na₂S). It is much more sensitive than the traditional Prussian Blue test.[5]

THE SODA LIME TEST

In a 10×75-mm test tube, mix about 50 mg of soda lime and 50 mg of MnO₂. Add 1 drop of a liquid unknown or about 10 mg of a solid unknown. Place over the mouth of the tube a moist strip of Brilliant Yellow paper (moist, red litmus paper is an alternative). Using a test tube holder, hold the tube at an incline (*pointing away from you and others*) and heat the contents gently at first and then quite strongly. Nitrogen-containing compounds will usually evolve ammonia.

A positive test for nitrogen is the deep-red coloration of the Brilliant Yellow paper (or blue color of the litmus paper).

The halogens (except fluorine)

USING THE FUSION SOLUTION

In a 10×75-mm test tube containing a boiling stone, place 0.5 mL (calibrated Pasteur pipet) of the fusion solution. Carefully acidify this solution by the dropwise addition of dilute HNO₃ (nitric acid), delivered from a Pasteur pipet (test acidity with litmus paper). If a positive test for nitrogen or sulfur was obtained, heat the resulting solution to a gentle boil (stir with a microspatula to prevent boilover) for 1 min over a microburner in the **hood** to expel any HCN or H₂S that might be present. Then cool the tube to room temperature.

To the resulting cooled solution, add 2 drops (Pasteur pipet) of aqueous 0.1 M AgNO₃ solution.

A heavy curdy-type precipitate is a positive test for the presence of Cl⁻, Br⁻, or I⁻ ion. A faint turbidity is a negative test.

AgCl precipitate is white.

AgBr precipitate is pale yellow.

AgI precipitate is yellow.

AgF is not detected by this test since it is relatively soluble in water.

The silver halides have different solubilities in dilute ammonium hydroxide solution.

Centrifuge the test tube and contents and remove the supernatant liquid using a Pasteur filter pipet. Add 0.5 mL (calibrated Pasteur pipet) of dilute ammonium hydroxide solution to the precipitate and stir with a glass rod to determine whether the solid is soluble.

---

[5]See Vogel, A. I. *Elementary Practical Organic Chemistry*, Part 2, 2nd ed.; Wiley: New York, 1966, p. 37.

AgCl is soluble in ammonium hydroxide due to the formation of the complex ion, $[Ag(NH_3)_2]^+$.

AgBr is slightly soluble in this solution.

AgI is insoluble in this solution.

A FURTHER TEST

Once the presence of a halide ion has been established, a further test is available to help you distinguish between Cl^-, Br^-, and I^- ions.[6]

As described above, acidify 0.5 mL of the fusion solution with dilute HNO_3. To this solution, add 5 drops (Pasteur pipet) of a 1.0% aqueous $KMnO_4$ solution and shake the test tube for about 1 min.

Now add 10–15 mg of oxalic acid, enough to decolorize the excess purple permanganate, followed by 0.5 mL of methylene chloride solvent. Stopper, shake, and vent the test tube and allow the layers to separate. Observe the color of the CH_2Cl_2 (lower) layer.

A clear methylene chloride layer indicates Cl^- ion.

A brown methylene chloride layer indicates Br^- ion.

A purple methylene chloride layer indicates I^- ion.

The colors may be faint and should be observed against a white background.

THE BEILSTEIN TEST[7]

In the Beilstein test organic compounds that contain chlorine, bromine, or iodine, and hydrogen decompose on ignition in the presence of copper oxide, to yield the corresponding hydrogen halides. These hydrogen halides react to form the volatile cupric halides that impart a green or blue-green color to a nonluminous flame. It is a very sensitive test, but some nitrogen-containing compounds and some carboxylic acids also give positive results.

Pound one end of a 4-in. long copper wire to form a flat surface that can act as a spatula. The other end of the wire is stuck in a cork stopper to serve as an insulated handle.

Heat the flat tip of the wire in a flame until coloration of the flame is negligible.

On the **cooled** flat surface of the wire, place a drop (Pasteur pipet) of liquid unknown or a few milligrams of solid unknown. Gently heat the material in the flame. The carbon present in the compound will burn first, so the flame will be luminous, but then the characteristic green or blue-green color may be evident. It may be fleeting, so watch carefully.

It is recommended that a known compound containing a halogen be tested so that you become familiar with the appearance of the expected color.

Fluoride ion is not detected by this test, since copper fluoride is not volatile.

---

[6]For a further test to distinguish between the three halide ions see Shriner, R. L.; Fuson, R. C.; Morrill, T. C. *The Systematic Identification of Organic Compounds*, 6th ed.; Wiley: New York, 1980, p. 81. Also see this reference (p. 85) for a specific test for the F^- ion.

[7]Beilstein, F. *Berichte* **1872**, 5, 620.

SOLUBILITY CHARACTERISTICS

Determination of the solubility characteristics of an organic compound can often give valuable information as to its structural composition. It is especially useful when correlated with spectral analysis. Several schemes have been proposed that place a substance in a definite group according to its solubility in various solvents. The scheme presented below is similar to that outlined in Shriner et al.[8]

There is no sharp dividing line between soluble and insoluble, and an arbitrary ratio of solute to solvent must be selected. We suggest that a compound be classified as soluble if its solubility is greater than 15 mg/500 μL of solvent.

Carry out the solubility determinations, at ambient temperature, in 10 × 75-mm test tubes. Place the sample (~15 mg) in the test tube and add a total of 0.5 mL of solvent in three portions from a graduated or calibrated Pasteur pipet. Between the addition of each portion, stir the sample vigorously with a glass stirring rod for 1.5–2 min. If the sample is water soluble, test the solution with litmus paper to assist in classification according to the solubility scheme that follows.

Note

> To test with litmus paper, dip the end of a small glass rod into the solution and then gently touch the litmus paper with the rod. **Do not dip the litmus paper into the test solution.**

In doing the solubility tests follow the scheme in the order given. *Keep a record of your observations.*

Step I. Test for water solubility. If soluble, test the solution with litmus paper.

Step II. If water soluble, determine the solubility in diethyl ether. This test further classifies water-soluble materials.

Step III. Water-insoluble compounds are tested for solubility in a 5% aqueous NaOH solution. If soluble, determine the solubility in 5% aqueous $NaHCO_3$. The use of the $NaHCO_3$ solution aids in distinguishing between strong (soluble) and weak (insoluble) acids.

Step IV. Compounds insoluble in 5% aqueous NaOH are tested for solubility in a 5% HCl solution.

Step V. Compounds insoluble in 5% aqueous HCl are tested with concentrated H_2SO_4. If soluble, further differentiation is made using 85% H_3PO_4, as shown in the scheme.

Step VI. Miscellaneous neutral compounds containing oxygen, sulfur, or nitrogen are normally soluble in strong acid solution.

---

[8]Shriner, R. L.; Fuson, R. C.; Morrill, T. C. *The Systematic Identification of Organic Compounds*, 6th ed.; Wiley: New York, 1980.

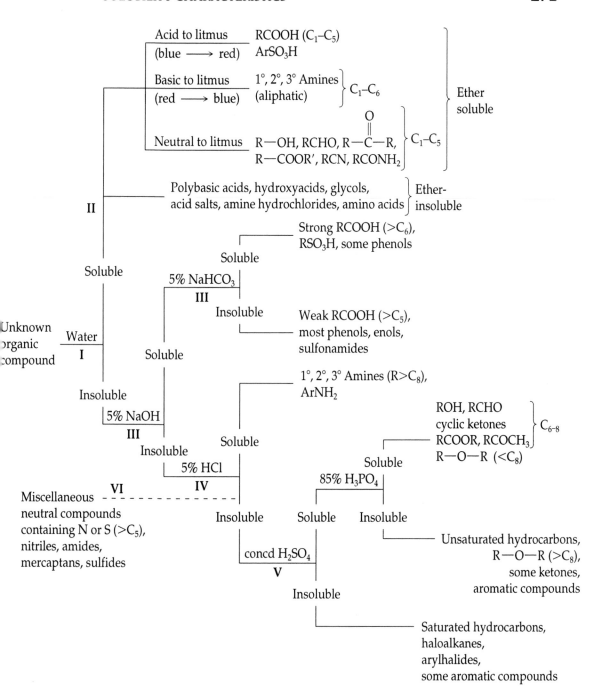

To classify a given compound, it may not be necessary to test its solubility in every solvent. *Do only those tests that are required to place the compound in one of the solubility groups.* Make your observations with care, and proceed in a logical sequence as you make the tests.

THE CLASSIFICATION TESTS[9]

Note

For all tests given in this section, drops of reagents are measured out using Pasteur pipets.

Alcohols

CERIC NITRATE TEST

INSTRUCTOR PREPARATION. The reagent is prepared by dissolving 4.0 g of ceric ammonium nitrate $[(NH_4)_2Ce(NO_3)_6]$ in 10 mL of 2 M HNO_3. Warming may be necessary.

Primary, secondary, and tertiary alcohols with fewer than 10 carbon atoms give a positive test as indicated by a change in color from *yellow* to *red*:

$$(NH_4)_2Ce(NO_3)_6 + RCH_2OH \longrightarrow [\text{alcohol} + \text{reagent}]$$
Yellow (Red complex)

Place 5 drops of test reagent on a white spot plate. Add 1–2 drops of the unknown sample (5 mg if a solid). Stir with a thin glass rod to mix the components and observe any color change.

1. If the alcohol is water insoluble, 3–5 drops of dioxane may be added, but run a blank to make sure the dioxane is pure. Efficient stirring gives positive results with most alcohols.

2. Phenols, if present, give a brown color or precipitate.

CHROMIC ANHYDRIDE TEST: THE JONES OXIDATION

INSTRUCTOR PREPARATION. The reagent is prepared by slowly adding a suspension of 1.0 g of CrO_3 in 1.0 mL of concentrated H_2SO_4 to 3 mL of water. Allow the solution to cool to room temperature before using.

The Jones oxidation test is a rapid method to distinguish primary and secondary alcohols from tertiary alcohols. A positive test is indicated by a color change from *orange* (the oxidizing agent, Cr^{6+}) while the oxidizing agent is itself reduced to the *blue green* (Cr^{3+}):

$$\left.\begin{array}{c} RCH_2OH \\ \text{or} \\ R_2CHOH \end{array}\right\} + H_2Cr_2O_7 \xrightarrow{H_2SO_4} Cr_2(SO_4)_3 + \begin{array}{c} RCO_2H \\ \text{or} \\ R_2C{=}O \end{array}$$
Orange   Green

[9]For a detailed discussion of classification tests see (a) Shriner, R. L.; Fuson, R. C.; Morrill, T. C. *The Systematic Identification of Organic Compounds*, 6th ed.; Wiley: New York, 1980, p. 138; (b) Pasto, D. J.; Johnson, C. R.; Miller, M. J. *Experiments and Techniques in Organic Chemistry*; Prentice Hall: Englewood Cliffs, NJ, 1992.

The test is based on oxidation of a primary alcohol to an aldehyde or acid, and of a secondary alcohol to a ketone.

On a white spot plate, place 1 drop of the liquid unknown (10 mg if a solid). Add 10 drops of acetone and stir the mixture with a thin glass rod. Add 1 drop of the test reagent to the resulting solution. Stir and observe any color change within a 2-second time period.

1. Run a blank to make sure the acetone is pure.

2. Tertiary alcohols, unsaturated hydrocarbons, amines, ethers, and ketones give a negative test within the 2-s time frame for observing the color change. Aldehydes, however, give a positive test, since they are oxidized to the corresponding carboxylic acids.

THE HCl/ZnCl$_2$ TEST: THE LUCAS TEST

INSTRUCTOR PREPARATION. The Lucas reagent is prepared by dissolving 16 g of anhydrous ZnCl$_2$ in 10 mL of concd HCl while it is cooling in an ice bath.

The Lucas test is used to distinguish between primary, secondary, and tertiary monofunctional alcohols having fewer than six carbon atoms:

$$R\text{---}OH + H^+ \xrightarrow{\text{ZnCl}_2} R^+ + H_2O$$

Soluble

$$R^+ \xrightarrow{Cl^-} RCl$$

Insoluble

The test requires that the alcohol initially be soluble in the Lucas test reagent solution. As the reaction proceeds, the corresponding alkyl chloride is formed, which is insoluble in the reaction mixture. As a result, the solution becomes cloudy. In some cases a separate layer may be observed.

1. Tertiary, allyl, and benzyl alcohols react to give an immediate cloudiness to the solution. You may be able to see a separate layer of the alkyl chloride after a short time.

2. Secondary alcohols generally produce a cloudiness within 3–10 min. The solution may have to be heated to obtain a positive test.

3. Primary alcohols having less than six carbon atoms dissolve in the reagent but react very, very slowly. Those having more than six carbon atoms do not dissolve to any significant extent, no reaction occurs, and the aqueous phase remains clear.

4. A further test to aid in distinguishing between tertiary and secondary alcohols is to run the test using concentrated hydrochloric acid. Tertiary alcohols react immediately to give the corresponding alkyl halide, whereas secondary alcohols do not react under these conditions.

In a small test tube prepared by sealing a Pasteur pipet off at the shoulder (see diagram, next page), place 2 drops of the unknown (10 mg if a solid) followed by 10 drops of the Lucas reagent.

Shake or stir the mixture with a thin glass rod and allow the solution to stand. Observe the results. Based on the times just given, classify the alcohol.

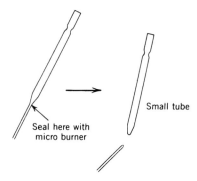

Seal here with
micro burner

Small tube

ADDITIONAL POINTS TO CONSIDER:

1. Certain polyfunctional alcohols also give a positive test.

2. If an alcohol having three or fewer carbons is expected, a 1-mL conical vial equipped with an air condenser should be used to prevent low molecular weight alkyl chlorides (volatile) from escaping and thus remaining undetected.

THE IODOFORM TEST

This test is positive for compounds that generate methyl ketones (or acetaldehyde) under the oxidation reaction conditions. For example, methyl carbinols (secondary alcohols having at least one methyl group attached to the carbon atom to which the —OH is attached), acetaldehyde, and ethanol give positive results.

For the test see Methyl Ketones and Methyl Carbinols (p. 286).

Periodic acid: vicinal diols

INSTRUCTOR PREPARATION. This reagent solution is prepared by dissolving 250 mg of periodic acid (H_5IO_6) in 50 mL of deionized water.

Vicinal diols (1,2 diols) are differentiated from the simple alcohols by the characteristic reaction below. Metaperiodic acid (HIO_4) selectively oxidizes 1,2-diols to give carbonyl compounds:

$$\underset{\text{1,2-Diol}}{\overset{\ddot{O}H \quad \ddot{O}H}{\underset{|}{\overset{|}{C}}-\underset{|}{\overset{|}{C}}}} \xrightarrow[\text{H}_2\text{O}]{\text{HIO}_4} 2 \; \underset{/}{\overset{\backslash}{C}}=\ddot{\ddot{O}} + H_2O + HIO_3$$

The test is based on the *instantaneous* formation of a white precipitate of silver iodate ($AgIO_3$) following addition of silver nitrate:

$$HIO_3 + AgNO_3 \longrightarrow HNO_3 + AgIO_3 \downarrow$$

Place 2 mL of the periodic acid reagent solution in a small test tube.
Add 2 drops of concentrated nitric acid and mix the solution thoroughly.
Add 2 drops of a liquid unknown (~2–5 mg of a solid) and mix again. Now
add 2–3 drops of 5% aqueous silver nitrate solution. An *instantaneous white precipitate* constitutes a positive test.

α-Hydroxyaldehydes, α-hydroxyketones, α-hydroxyacids, 1,2-diketones,
and α-aminoalcohols also give a positive test.

Aldehydes and ketones

THE 2,4-DINITROPHENYLHYDRAZINE TEST

*INSTRUCTOR PREPARATION. The reagent solution is prepared by dissolving
1.0 g of 2,4-dinitrophenylhydrazine in 5.0 mL of concentrated sulfuric acid. This solution is slowly added, with stirring, to a mixture of 10 mL of water and 35 mL of
95% ethanol. After mixing, filter the solution.*

Aldehydes and ketones react rapidly with 2,4-dinitrophenylhydrazine to
form 2,4-dinitrophenylhydrazones. These derivatives range in color from *yellow*
to *red*, depending on the degree of conjugation in the carbonyl compound:

2,4-Dinitrophenylhydrazine   Yellow-to-red precipitate

On a white spot plate place 7–8 drops of 2,4-dinitrophenylhydrazine
reagent solution.

Then add 1 drop of a liquid unknown. If the unknown is a solid, add 1
drop of a solution prepared by dissolving 10 mg of the material in 10 drops of
ethanol. The mixture is stirred with a thin glass rod. The formation of a red-
to-yellow precipitate is a positive test.

Note

> The reagent, 2,4-dinitrophenylhydrazine, is orange-red and melts at
> 198 °C (dec). Do not mistake it for a derivative!

Reactive esters or anhydrides react with the reagent to give a positive test.
Allylic or benzylic alcohols may be oxidized to aldehydes or ketones, which in
turn give a positive result. Amides do not interfere with the test. Be sure that
your unknown is pure and does not contain aldehyde or ketone impurities.

Phenylhydrazine and *p*-nitrophenylhydrazine are often used to prepare the corresponding hydrazones. These reagents also yield solid derivatives of aldehydes and ketones.

SILVER MIRROR TEST FOR ALDEHYDES: TOLLENS REAGENT

This reaction involves the oxidation of aldehydes to the corresponding carboxylic acid, using an alcoholic solution of silver ammonium hydroxide. A positive test is the formation of a *silver* mirror, or a black precipitate of finely divided silver:

$$RC=\ddot{O} + 2\,Ag(NH_3)_2\ddot{O}H \longrightarrow 2\,Ag\downarrow + R-C\overset{\ddot{O}:}{\underset{\ddot{O}:^-,\,NH_4^+}{\diagdown}} + H_2\ddot{O} + 3\,\ddot{N}H_3$$

The test should be run only after the presence of an aldehyde or ketone has been established.

In a small test tube prepared from a Pasteur pipet (see the Lucas test), place 1.0 mL of a 5% aqueous solution of $AgNO_3$, followed by 1 drop of aqueous 10% NaOH solution. Now add concentrated aqueous ammonia, drop by drop (2–4 drops) with shaking, until the precipitate of silver oxide just dissolves. Add 1 drop of the unknown (10 mg if a solid), with shaking, and allow the reaction mixture to stand for 10 min at room temperature. If no reaction has occurred, place the test tube in a sand bath at 40 °C for 5 min. Observe the result.

ADDITIONAL POINTS TO CONSIDER:

1. Avoid a large excess of ammonia.

2. Reagents must be well mixed. Stirring with a thin glass rod is recommended.

3. *This reagent is freshly prepared for each test. It should not be stored since decomposition occurs with the formation of AgN_3, which is explosive.*

4. This oxidizing agent is very mild, and thus alcohols are not oxidized under these conditions. Ketones do not react. Some sugars, acyloins, hydroxylamines, and substituted phenols do give a positive test.

CHROMIC ACID TEST

INSTRUCTOR PREPARATION. The reagent is prepared by dissolving 1 g of chromium trioxide in 1 mL of concd H_2SO_4, followed by 3 mL of H_2O.

Chromic acid in acetone rapidly oxidizes aldehydes to carboxylic acids. Ketones react very slowly, or not at all.

In a 3-mL vial or small test tube, place 2 drops of a liquid unknown (~10 mg if a solid) and 1 mL of spectral-grade acetone. Now add several drops of the chromic acid reagent.

A green precipitate of chromous salts is a positive test. Aliphatic aldehydes give a precipitate within 30 s; aromatic aldehydes take 30–90 s.

The reagent also reacts with primary and secondary alcohols (see Chromic Anhydride Test: Jones Oxidation, p. 272).

BISULFITE ADDITION COMPLEXES

INSTRUCTOR PREPARATION. The reagent is prepared by mixing 1.5 mL of ethanol and 6 mL of a 40% aqueous solution of sodium bisulfite. Filter the reagent before use, if a small amount of the salt does not dissolve.

Most aldehydes react with a saturated sodium bisulfite solution to yield a crystalline bisulfite addition complex:

$$\overset{\backslash}{\underset{/}{C}}=\ddot{\underset{\cdot\cdot}{O}} \quad \underset{H^+ \text{ or } HO^-}{\overset{NaHSO_3}{\rightleftharpoons}} \quad \overset{\backslash}{\underset{/}{C}}\overset{SO_3^-, Na^+}{\underset{\ddot{\underset{\cdot\cdot}{O}}H}{\Big|}}$$

The reaction is reversible and thus the carbonyl compound can be recovered by treatment of the complex with aqueous 10% $NaHCO_3$ or dilute HCl solution.

Place 50–75 µL of the liquid unknown in a small test tube and add 150 µL of the sulfite reagent and mix thoroughly.

A crystalline precipitate is a positive test.

Alkyl methyl ketones and unhindered cyclic ketones also give a positive test.

Alkanes and cycloalkanes: Saturated hydrocarbons

IODINE CHARGE-TRANSFER COMPLEX

Alkanes exhibit a *negative* iodine charge-transfer complex test. Species containing π electrons or nonbonded electron pairs produce a brown solution. This color formation is due to the charge-transfer complex between iodine and the available electrons:

$$\overset{\backslash/}{\underset{\backslash}{\underset{/}{\overset{C}{\underset{C}{\|}}}}}\cdots I_2 \quad \text{or} \quad \overset{\backslash}{\underset{/}{\ddot{O}}}\colon\cdots I_2$$

Solutions of iodine and nonparticipating compounds are violet in color.

On a white spot plate, place a small crystal of iodine. Now add 2–3 drops of a liquid unknown. Alkanes give a *negative* test (violet color).

The test is run only on liquid unknowns. Saturated hydrocarbons, fluorinated and chlorinated saturated hydrocarbons, and aromatic hydrocarbons and their halogenated derivatives all give violet solutions. All other species give a positive test (brown solution).

CONCENTRATED SULFURIC ACID

Saturated hydrocarbons, halogenated saturated hydrocarbons, simple aromatic hydrocarbons, and their halogenated derivatives are insoluble in *cold* concentrated sulfuric acid.

In a small test tube, using **caution,** place 100 μL of *cold* concentrated sulfuric acid. Now add 50 μL of an unknown. A resulting heterogeneous solution (the unknown does *not* dissolve) is a positive test for a saturated hydrocarbon.

Alkenes, and compounds having a functional group containing a nitrogen or oxygen atom, are soluble in cold, concentrated acid.

Alkenes and alkynes: Unsaturated hydrocarbons

BROMINE IN METHYLENE CHLORIDE
Unsaturated hydrocarbons readily add bromine (Br_2):

The test is based on the decolorization of a red-brown bromine-methylene chloride solution.

Caution

Bromine is highly toxic and can cause burns.

In a 10×75-mm test tube, or in a small tube prepared from a Pasteur pipet (see Lucas test), place 2 drops of a liquid unknown (~15 mg if a solid) followed by 0.5 mL of methylene chloride. Add dropwise, in the **hood** with shaking, a 2% solution of bromine in methylene chloride solvent. The presence of an unsaturated hydrocarbon will require 2–3 drops of the reagent before the reddish-brown color of bromine persists in the solution.

ADDITIONAL POINTS TO CONSIDER:

1. Methylene chloride is used in place of the usual carbon tetrachloride (CCl_4) because it is less toxic.

2. Phenols, enols, amines, aldehydes, and ketones interfere with this test.

PERMANGANATE TEST: BAEYER TEST
FOR UNSATURATION
Unsaturation in an organic compound can be detected by the decolorization of permanganate solution. The reaction involves the cis hydroxylation of the alkene to give a 1,2 diol (glycol):

On a white spot plate, place 0.5 mL of *alcohol-free* acetone, followed by 2 drops of the unknown compound ($\sim$ 15 mg if a solid). Now add dropwise (2–3 drops), with stirring, a 1% aqueous solution of potassium permanganate ($KMnO_4$). A positive test for unsaturation is the discharge of purple permanganate color from the reagent and the precipitation of brown manganese oxides.

Any functional group that undergoes oxidation with permanganate interferes with the test (phenols, aryl amines, most aldehydes, primary and secondary alcohols, etc.).

Alkyl halides

SILVER NITRATE TEST

Alkyl halides that undergo the S_N1 substitution reaction react with alcoholic silver nitrate ($AgNO_3$) to form a precipitate of the corresponding silver halide.

Secondary and primary halides react slowly or not at all at room temperature. However, they do react at elevated temperatures. Tertiary halides react immediately at room temperature.

In a 1.0-mL conical vial place 0.5 mL of 2% ethanolic $AgNO_3$ solution and 1 drop of unknown ($\sim$10 mg if a solid). A positive test is indicated by the appearance of a precipitate within 5 min. If no reaction occurs, add a boiling stone and equip the vial with an air condenser. Heat the solution at *gentle* reflux for an additional 5 min using a sand bath. Cool the solution.

If a precipitate is formed, add 2 drops of dilute HNO_3. Silver halides will not dissolve in nitric acid solution.

ADDITIONAL POINTS TO CONSIDER:

1. The order of reactivity for R groups is allyl $\cong$ benzyl > tertiary > secondary >>> primary. For the halide leaving groups the order is I > Br > Cl.

2. Acid halides, α-haloethers, and 1,2-dibromo compounds also give a positive test at room temperature. Only activated aryl halides give a positive test at elevated temperatures.

SODIUM IODIDE IN ACETONE

INSTRUCTOR PREPARATION. The reagent is prepared by dissolving 3 g of sodium iodide (NaI) in 25 mL of acetone. Store in a dark bottle.

Primary alkyl chlorides and bromides can be distinguished from aryl and alkenyl halides by reaction with sodium iodide in acetone (Finkelstein reaction):

$$R\text{—}X + NaI \xrightarrow{\text{acetone}} R\text{—}I + NaX \downarrow$$
$$X = Cl, Br$$

Primary alkyl bromides undergo an S_N2 displacement reaction within 5 min at room temperature, and primary alkyl chlorides only at 50 °C.

In a 1.0-mL conical vial, place 1 drop of a liquid unknown ($\sim$10 mg if a solid) and 3 drops of acetone. To this solution add 0.5 mL of sodium iodide–acetone reagent.

A positive test is the appearance of a precipitate of NaX within 5 min. If no precipitate is observed, add a boiling stone and equip the vial with an air condenser. Warm the reaction mixture in a sand bath at about 50 °C for 5 min. Cool to room temperature and determine whether a reaction has occurred.

ADDITIONAL POINTS TO CONSIDER:

1. Benzylic and allylic chlorides and bromides, acid chlorides and bromides, and α-haloketones, α-haloesters, α-haloamides, and α-halonitriles also give a positive test at room temperature.

2. Primary and secondary alkyl chlorides, and secondary and tertiary alkyl bromides, react at 50 °C under these conditions.

3. If the solution turns red brown in color, iodine is being liberated.

Amides, ammonium salts, and nitriles

HYDROXAMATE TEST FOR AMIDES
Unsubstituted (on nitrogen) amides, and the majority of substituted amides, will give a positive hydroxamate test:

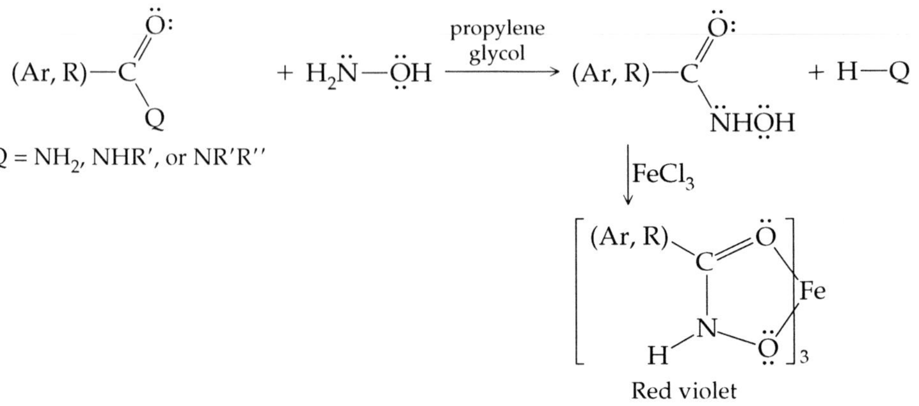

where Q = NH$_2$, NHR', or NR'R''

Red violet

The hydroxamic acid is identified by formation of a red-to-purple color in the presence of Fe^{3+} ion, as for the test with esters (see below).

In a 3.0-mL conical vial containing a boiling stone and equipped with an air condenser place 1 drop of a liquid unknown (~10 mg if a solid), followed by 0.5 mL of 1 M hydroxylamine hydrochloride–propylene glycol solution. Heat the resulting mixture to reflux temperature (~190 °C) using a sand bath, and reflux for 3–5 min. Cool the solution to room temperature, and add 2 drops of 5% aqueous FeCl$_3$ solution. The formation of a red-to-purple color is a positive test.

ALKALINE HYDROLYSIS
Ammonium salts, amides, and nitriles undergo hydrolysis in alkaline solution to form ammonia gas, or an amine:

$$R-\overset{\overset{\cdot\cdot}{O}\cdot}{\underset{\parallel}{C}}-\overset{\cdot\cdot}{N}H_2 \xrightarrow[\text{H}_2\text{O}]{\text{NaOH}} R-\overset{\overset{\cdot\cdot}{O}:}{C}\diagdown_{\underset{\cdot\cdot}{O}\overset{\cdot\cdot}{:},\,Na^+} + \overset{\cdot\cdot}{N}H_3\uparrow$$

$$R-\overset{\overset{\cdot\cdot}{O}\cdot}{\underset{\parallel}{C}}-\overset{\cdot\cdot}{N}HR' \xrightarrow[\text{H}_2\text{O}]{\text{NaOH}} R-\overset{\overset{\cdot\cdot}{O}:}{C}\diagdown_{\underset{\cdot\cdot}{O}\overset{\cdot\cdot}{:},\,Na^+} + H_2\overset{\cdot\cdot}{N}R'\uparrow$$

$$R-\overset{\overset{\cdot\cdot}{O}\cdot}{\underset{\parallel}{C}}-\overset{\cdot\cdot}{N}R'_2 \xrightarrow[\text{H}_2\text{O}]{\text{NaOH}} R-\overset{\overset{\cdot\cdot}{O}:}{C}\diagdown_{\underset{\cdot\cdot}{O}\overset{\cdot\cdot}{:},\,Na^+} + H\overset{\cdot\cdot}{N}R'_2\uparrow$$

$$R-CN: \xrightarrow[\text{H}_2\text{O}]{\text{NaOH}} R-\overset{\overset{\cdot\cdot}{O}:}{C}\diagdown_{\underset{\cdot\cdot}{O}\overset{\cdot\cdot}{:},\,Na^+} + \overset{\cdot\cdot}{N}H_3\uparrow$$

Detection of ammonia from ammonium salts, primary amides, and nitriles, by use of a color test using copper sulfate solution, constitutes a positive test for these functional groups. The same test may also be used for secondary and tertiary amides that can generate low molecular weight (volatile) amines upon hydrolysis.

In a 1.0-mL conical vial containing a boiling stone, and equipped with an air condenser, place 1–2 drops of the unknown liquid (~10 mg if a solid) and 0.5 mL of 20% aqueous NaOH solution. Heat this mixture to *gentle* reflux on a sand bath. Moisten a strip of filter paper with 2 drops of 10% aqueous copper sulfate solution and place it over the top of the condenser. Formation of a *blue* color (copper ammonia [or amine] complex) is a positive test.

The filter paper may be held in place using a small test tube holder or other suitable device.

Amines

COPPER ION TEST
Amines will give a blue-green coloration or precipitate when added to a copper sulfate solution. In a small test tube, place 0.5 mL of a 10% copper sulfate solution. Now add 1 drop of an unknown (~10 mg if a solid). The blue-green coloration or precipitate is a positive test. Ammonia will also give a positive test.

HINSBERG TEST
The Hinsberg test is useful for distinguishing between primary, secondary, and tertiary amines. The reagent used is *p*-toluenesulfonyl chloride in alkaline solution.

Primary amines with fewer than seven carbon atoms form a sulfonamide that is soluble in the alkaline solution. Acidification of the solution results in the precipitation of the insoluble sulfonamide:

$$H_3C-\langle\rangle-SO_2Cl + R-NH_2 \xrightarrow{NaOH} H_3C-\langle\rangle-SO_2\bar{N}R, Na^+$$

$$H_3C-\langle\rangle-SO_2\bar{N}R, Na^+ \underset{\substack{excess \\ base}}{\overset{\substack{excess \\ acid}}{\rightleftharpoons}} H_3C-\langle\rangle-SO_2NHR + NaCl + H_2O$$

(soluble)   (insoluble)

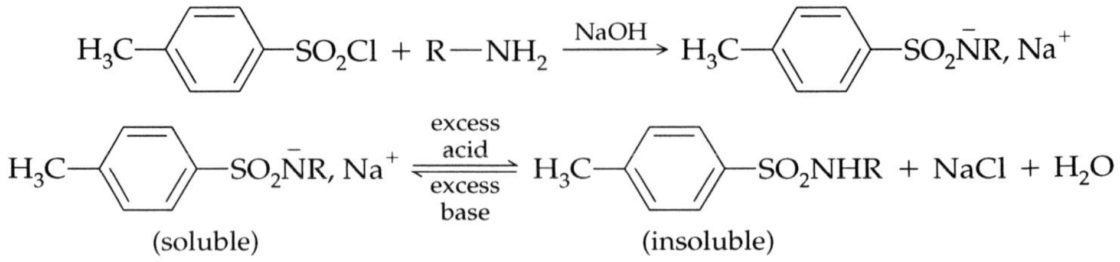

Secondary amines form an insoluble sulfonamide in the alkaline solution:

$$H_3C-\langle\rangle-SO_2Cl + R_2NH \xrightarrow{NaOH}$$

$$H_3C-\langle\rangle-SO_2NR_2 + NaCl + H_2O \xrightarrow{\substack{excess \\ base}} \text{no change}$$

(insoluble)

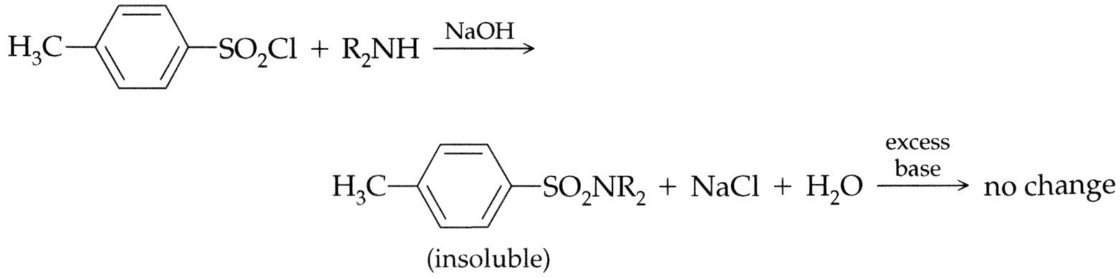

Tertiary amines normally give no reaction under these conditions:

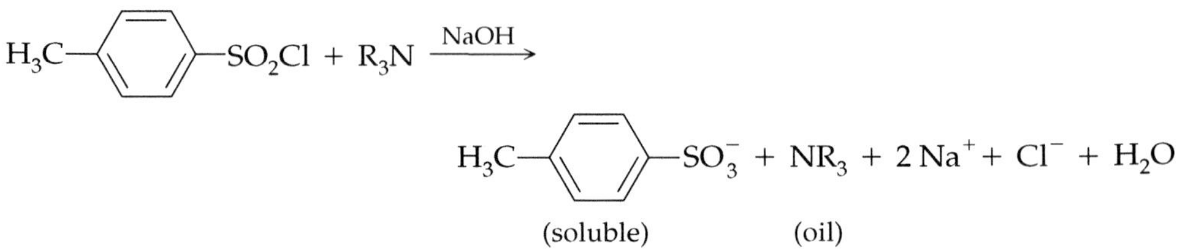

$$H_3C-\langle\rangle-SO_2Cl + R_3N \xrightarrow{NaOH}$$

$$H_3C-\langle\rangle-SO_3^- + NR_3 + 2\,Na^+ + Cl^- + H_2O$$

(soluble)   (oil)

In a 1.0-mL conical vial containing a boiling stone, and equipped with an air condenser, place 0.5 mL of 10% aqueous sodium hydroxide solution, 1 drop of the sample unknown (~10 mg if a solid), followed by 30 mg of *p*-toluene-sulfonyl chloride (in the **hood**). Heat the mixture to reflux for 2–3 min on a sand bath, and then cool it in an ice bath. Test the alkalinity of the solution using litmus paper. If it is not alkaline, add additional 10% aqueous sodium hydroxide dropwise.

Using a Pasteur filter pipet, separate the solution from any solid that may be present. Transfer the solution to a clean 1.0-mL conical vial and **save.**

Note

If an oily upper layer is obtained at this stage, remove the lower alkaline phase using a Pasteur filter pipet and **save.** To the remaining oil add 0.5 mL of cold water and stir vigorously to obtain a solid material.

If a solid is obtained, it may be (1) the sulfonamide of a secondary amine; (2) recovered tertiary amine, if the original amine was a solid; or (3) the insoluble salt of a primary sulfonamide derivative, if the original amine had more than six carbon atoms.

ADDITIONAL POINTS TO CONSIDER:

1. If the solid is a tertiary amine, it is soluble in aqueous 10% HCl.

2. If the solid is a secondary sulfonamide, it is insoluble in aqueous 10% NaOH.

3. If no solid is present, acidify the alkaline solution by adding 10% aqueous HCl. If the unknown amine is primary, the sulfonamide will precipitate.

BROMINE WATER

Aromatic amines, since they possess an electron-rich aromatic ring, can undergo electrophilic aromatic substitution with bromine, to yield the corresponding arylamino halide(s). Therefore, if elemental tests indicate that an aromatic group is present in an amine, treatment with the bromine water reagent may indicate that the amine is attached to an aromatic ring.

For the test, see Phenols and Enols (p. 288).

Aromatic hydrocarbons with no functional groups

FUMING SULFURIC ACID

Simple aromatic hydrocarbons are insoluble in sulfuric acid (H_2SO_4) but are soluble in fuming sulfuric acid. If these hydrocarbons contain more than two alkyl substituents, they may be sulfonated under these conditions.

In a small test tube place 100 μL of fuming sulfuric acid, using **caution.** Now add 50 μL of the unknown suspected to be aromatic. A resulting homogeneous solution is a positive test.

AZOXYBENZENE AND ALUMINUM CHLORIDE

This color test is run only on those aromatic compounds that are insoluble in sulfuric acid (see previous test). The color produced in this test results from the formation of a complex of $AlCl_3$ and a *p*-arylazobenzene derivative:

Azoxybenzene   Colored complex

In a small dry test tube, place 250 μL of the aromatic unknown. Add a small crystal of azoxybenzene and about 12 mg of anhydrous aluminum chloride. If a color is not produced immediately, warm the mixture for a few minutes.

Aryl halides and other simple aromatic hydrocarbons give a deep-orange to dark-red color or precipitate. Polynuclear aromatic hydrocarbons, such as naphthalenes and anthracenes, give brown colors. Aliphatic hydrocarbons give no color, or at most a light-yellow tint.

Carboxylic acids

The presence of a carboxylic acid is detected by its solubility behavior. An aqueous solution of the acid will be acidic to litmus paper (or pH paper may be used). Since a sulfonic acid would also give a positive test, the test for sulfur (sodium fusion) is used to distinguish between the two types of acids. A water-soluble phenol is acidic toward litmus paper but also would give a positive ferric chloride test.

Carboxylic acids also react with a 5% solution of sodium bicarbonate.

Place 1–2 mL of the bicarbonate solution on a watch glass and add 1–2 drops of the acid (~10 mg if a solid). Gas bubbles of CO_2 constitute a positive test.

Esters

HYDROXAMATE TEST

Carboxylic esters can be identified by conversion to hydroxamic acid salts. Acidification of this salt produces the corresponding hydroxamic acid (RCONHOH), which is identified by formation of a red-to-purple color in the presence of Fe^{3+} ion:

Red violet

In a 3.0-mL conical vial containing a boiling stone, and equipped with an air condenser, place 1 drop of the liquid unknown (~10 mg if a solid) followed by 0.5 mL of 1.0 M ethanolic hydroxylamine hydrochloride solution. Add 10% methanolic KOH to this solution (dropwise) until the resulting solution has pH ~10 (pH paper). Heat this mixture to reflux temperature using a sand bath for 5 min, cool to room temperature, and acidify to pH = 3–4 by dropwise addition of 5% aqueous HCl solution. Now add 2 drops of 5% aqueous $FeCl_3$ solution. The formation of a red-to-purple color is a positive test.

ADDITIONAL POINTS TO CONSIDER:

1. It is suggested that a blank be run for comparison purposes.

2. Acid chlorides, anhydrides, lactones, and imides also give a positive test.

SAPONIFICATION

This well-known reaction of esters can often be used to classify these compounds. It also may lead to a useful derivative if the corresponding carboxylic acid is isolated.

In a 3.0-mL conical vial containing a magnetic spin vane, place 100 μL of the liquid unknown (~150 mg if a solid) and add 1 mL of 6 M NaOH solution. Attach the vial to a reflux condenser. Now place the vial in a sand bath on a magnetic stirring hot plate and, with stirring, heat the mixture at reflux for 0.5 h, or until the solution becomes homogeneous.

A positive test is the disappearance of the organic layer (if the original unknown was water insoluble) or the lack of the usually pleasant aroma of the unknown ester.

High-boiling esters (bp > 200 °C) are usually not saponified under these conditions due to their low solubility in the aqueous solvent.

Ethers

Caution

> Upon standing, ethers may form peroxides. Peroxides are very explosive. To test for the presence of these substances, use starch–iodide paper that has been moistened with 6 M HCl. Peroxides cause the paper to turn blue. To remove peroxides from ethers, pass the material through a short column of highly activated alumina (Woelm basic alumina, activity grade 1).[10] Always retest for peroxides before using the ether.

FERROX TEST

The ferrox test is a color test sensitive to oxygen, which may be used to distinguish ethers from hydrocarbons that, like most ethers, are soluble in sulfuric acid.

In a dry 10 × 75-mm test tube using a glass stirring rod, grind a crystal of ferric ammonium sulfate and a crystal of potassium thiocyanate. The ferric hexathiocyanatoferrate that is formed adheres to the rod.

In a second clean 10 × 75-mm test tube, place 2–3 drops of a liquid unknown. If dealing with a solid, use about 10 mg and add toluene until a saturated solution is obtained. Now, using the rod with the ferric hexathiocyanatoferrate attached, stir the unknown. *If the unknown contains oxygen, the ferrate compound dissolves and a reddish-purple color is observed.*

Some high molecular weight ethers do not give a positive test.

BROMINE WATER

Since the aromatic ring is electron rich, aromatic ethers can undergo electrophilic aromatic substitution with bromine to yield the corresponding aryl ether–halide(s). Therefore, if elemental tests indicate that an aromatic group is

---

[10]Pasto, D. J.; Johnson, C. R.; Miller, M. J. *Experiments and Techniques in Organic Chemistry*; Prentice Hall: Englewood Cliffs, NJ, 1992, p. 33.

present in an ether, treatment with the bromine water reagent may substantiate the presence of an aryl ether.

For the test see Phenols and Enols (p. 288).

Methyl ketones and methyl carbinols

IODOFORM TEST

INSTRUCTOR PREPARATION. Dissolve 3 g of KI and 1 g I_2 in 20 mL of water.

The iodoform test involves hydrolysis and cleavage of methyl ketones to form a yellow precipitate of iodoform (CHI_3):

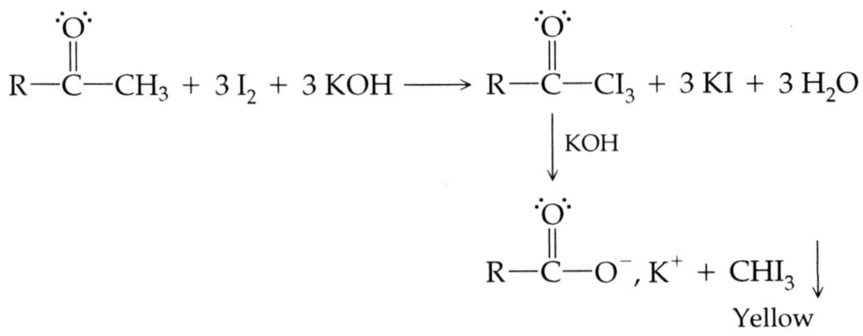

$$R-\overset{\overset{\ddot{O}}{\|}}{C}-CH_3 + 3\,I_2 + 3\,KOH \longrightarrow R-\overset{\overset{\ddot{O}}{\|}}{C}-CI_3 + 3\,KI + 3\,H_2O$$

$$\downarrow KOH$$

$$R-\overset{\overset{\ddot{O}}{\|}}{C}-O^-, K^+ + CHI_3 \downarrow$$

Yellow

It is also a positive test for compounds that, upon oxidation, generate methyl ketones (or acetaldehyde) under these reaction conditions. For example, methyl carbinols (secondary alcohols having at least one methyl group attached to the carbon atom to which the —OH unit is linked), acetaldehyde, and ethanol give positive results.

In a 3.0-mL conical vial equipped with an air condenser, place 2 drops of the unknown liquid (10 mg if a solid), followed by 5 drops of 10% aqueous KOH solution.

*Note*_____

> If the sample is insoluble in the aqueous phase, either mix vigorously or add dioxane (in the **hood**) or bis (2-methoxyethyl) ether to obtain a homogeneous solution.

Warm the mixture on a sand bath to 50–60 °C and add the KI–I_2 reagent dropwise until the solution becomes dark brown in color (~1.0 mL). Additional 10% aqueous KOH is now added (dropwise) until the solution is again colorless.

*Caution*_____

> Iodine is highly toxic and can cause burns.

After warming for 2 min, cool the solution and determine whether a yellow precipitate (CHI_3, iodoform) has formed. If a precipitate is not observed, reheat as before for another 2 min. Cool and check again for the appearance of iodoform.

AN ADDITIONAL POINT TO CONSIDER:

The iodoform test is reviewed elsewhere.[11]

Nitro compounds

FERROUS HYDROXIDE TEST

Many nitro compounds give a positive test based on the following reaction:

$$R{-}NO_2 + 4\,H_2O + 6\,Fe(OH)_2 \longrightarrow R{-}NH_2 + 6\,Fe(OH)_3 \downarrow$$
$$\text{Red-brown}$$

The nitro derivative oxidizes the iron(II) hydroxide to iron(III) hydroxide; the latter is a red-brown solid.

In a 1.0-mL conical vial place 5–10 mg of the unknown compound, followed by 0.4 mL of freshly prepared 5% aqueous ferrous ammonium sulfate solution. After mixing, add 1 drop of 3 M sulfuric acid followed by 10 drops of 2 M methanolic KOH. Cap the vial, shake vigorously, vent, and then allow it to stand over a 5-min period. The formation of a red-brown precipitate, usually within 1 min, is a positive test for a nitro group.

SODIUM HYDROXIDE COLOR TEST

Treatment of an aromatic nitro compound with 10% sodium hydroxide solution may often be used to determine the number of nitro groups present on the aromatic ring system.

Mononitro compounds produce no color (a light yellow may be observed).

Dinitro compounds produce a bluish-purple color.

Trinitro compounds produce a blood-red color.

The color formation is due to formation of Meisenheimer complexes (for a discussion, see Pasto et al.[12]).

To run the test, dissolve 10 mg of the unknown (1–2 drops if a liquid) in 1 mL of acetone in a small test tube. Now add about 200 μL of 10% NaOH solution and shake. Observe any color formation.

If amino, substituted amino, or hydroxyl groups are present in the molecule, a positive color test is not obtained.

---

[11]Fuson, R. C.; Bull, B. A. *Chem. Rev.* **1934**, *15*, 275.

[12]Pasto, D. J.; Johnson, C. R.; Miller, M. J. *Experiments and Techniques in Organic Chemistry*; Prentice Hall: Englewood Cliffs, NJ, 1992, p. 321.

Phenols and enols

FERRIC ION TEST

Most phenols and enols form colored complexes in the presence of ferric ion, Fe^{3+}:

$$6 \quad \underset{}{\text{C}_6\text{H}_5\text{OH}} + Fe^{3+} \rightleftharpoons \left[\left(\text{C}_6\text{H}_5\text{O} \right)_6 Fe \right]^{3-} + 6\,H^+$$

Phenols give red, blue, purple, or green colors. Sterically hindered phenols may give a negative test. Enols generally give a tan, red, or red-violet color.

On a white spot plate place 2 drops of water, or 1 drop of water plus 1 drop of ethanol, or 2 drops of ethanol, depending on the solubility characteristics of the unknown. To this solvent system add 1 drop (10 mg if a solid) of the substance to be tested. Stir the mixture with a thin glass rod to complete dissolution. Add 1 drop of 2.5% aqueous ferric chloride ($FeCl_3$) solution (light yellow in color). Stir and observe any color formation. If necessary, a second drop of the $FeCl_3$ solution may be added.

ADDITIONAL POINTS TO CONSIDER:

1. The color developed may be fleeting or it may last for many hours. A slight excess of the ferric chloride solution may or may not destroy the color.

2. An alternative procedure using $FeCl_3$–CCl_4 solution in the presence of pyridine is available.[13]

BROMINE WATER

Phenols, substituted phenols, aromatic ethers, and aromatic amines, since the aromatic rings are electron rich, undergo aromatic electrophilic substitution with bromine to yield substituted aryl halides. For example,

$$\text{C}_6\text{H}_5\text{OH} + 3\,Br_2 \xrightarrow{H_2O} \text{(2,4,6-tribromophenol)} + 3\,HBr$$

Caution

The test should be run in the **hood.**

---

[13]Soloway, S.; Wilen, S. H. *Anal. Chem.* **1952,** *4,* 979.

In a small test tube, place 1–2 drops of the unknown (~20 mg if a solid) and add 1–2 mL of water. Check the pH of the solution with pH paper. In the **hood,** add saturated bromine water dropwise until the bromine color persists. A precipitate generally forms.

A positive test is the decolorization of the bromine solution, and often the formation of an off-white precipitate. If the unknown is a phenol, this should cause the pH of the original solution to be less than 7.

PREPARATION OF DERIVATIVES

Based on the preliminary and classification tests carried out to this point, you should have established the type of functional group (or groups) present (or lack of one) in the unknown organic sample. The next step in qualitative organic analysis is to consult a set of tables containing a listing of known organic compounds sorted by functional group and/or by physical properties or by both. Using the physical-properties data for your compound, you can select a few possible candidates that appear to "fit" the data you have collected. On a chemical basis, the final step in the qualitative identification sequence is to prepare one or two *crystalline derivatives* of your compound. Selection of the specific compound, and thus final confirmation of its identity, can then be made from the extensive derivative tables that have been accumulated. With the advent of spectral analysis, the preparation of derivatives is often not necessary, but the wealth of chemistry that can be learned by the beginning student in carrying out these procedures is extensive and important. The preparation of selected derivatives for the most common functional groups are given next. Condensed tables of compounds and their derivatives are summarized on the website, ⚫ Appendix. For extensive tables and alternative derivatives that can be utilized, see the following Bibliography.

BIBLIOGRAPHY

PASTO, D. J.; JOHNSON, C. R.; MILLER, M. J.; *Experiments and Techniques in Organic Chemistry;* Prentice Hall: Englewood Cliffs, NJ, 1992.

RAPPOPORT, Z. *Handbook of Tables for Organic Compound Identification,* 3rd ed.; CRC Press: Boca Raton, FL, 1967.

SHRINER, R. L.; HERMANN, C. K. F.; MORRILL, T. C.; FUSON, R. C. *The Systematic Identification of Organic Compounds,* 7th ed.; Wiley: New York, 1998.

Note

In each of the procedures outlined below, drops of reagents are measured using Pasteur pipets.

CARBOXYLIC ACIDS[14]

Preparation of acid chlorides

$$R-\overset{\overset{\displaystyle \ddot{O}}{\|}}{C}-\ddot{O}H \ + \ Cl-\overset{\overset{\displaystyle \ddot{O}}{\|}}{S}-Cl \ \xrightarrow{\text{DMF}} \ R-\overset{\overset{\displaystyle \ddot{O}}{\|}}{C}-Cl \ + \ HCl\uparrow \ + \ SO_2\uparrow$$

Weigh and place 20 mg of the unknown acid in a dry 3.0-mL conical vial containing a boiling stone and fitted with a cap. Now, in the **hood**, add 4 drops of thionyl chloride and 1 drop of *N,N*-dimethylformamide (DMF). Immediately attach the vial to a reflux condenser that is protected by a calcium chloride drying tube.

Caution

This reaction is run in the **hood** since hydrogen chloride and sulfur dioxide are evolved. Thionyl chloride is an irritant and is harmful to breathe. Immediately recap the vial after each addition until the vial is attached to the reflux condenser.

Allow the mixture to stand at room temperature for 10 min, heat it at gentle reflux on a sand bath for 15 min, and then cool it to room temperature. Dilute the reaction mixture with 5 drops of methylene chloride solvent.

The acid chloride is not isolated but is used directly in the preparations that follow.

Amides

$$R-\overset{\overset{\displaystyle \ddot{O}}{\|}}{C}-Cl \ + \ 2\ \ddot{N}H_3 \ \longrightarrow \ R-\overset{\overset{\displaystyle \ddot{O}}{\|}}{C}-\ddot{N}H_2 \ + \ NH_4Cl$$

Cool the vial in an ice bath and add 10 drops of concentrated aqueous ammonia, in the **hood** via Pasteur pipet, *dropwise*, with stirring. *It is convenient to make this addition down the neck of the air condenser.* The amide may precipitate during this operation. After the addition is complete, remove the ice bath and stir the mixture for an additional 5 min. Now add methylene chloride (10 drops)

[14]See 🌐 Tables 10.1 and 10.2.

and stir the resulting mixture to dissolve any precipitate. Separate the methylene chloride layer from the aqueous layer using a Pasteur filter pipet and transfer it to another Pasteur filter pipet containing 200 mg of anhydrous sodium sulfate. Collect the eluate in a Craig tube containing a boiling stone. Extract the aqueous phase with an additional 0.5 mL of methylene chloride. Separate the methylene chloride layer as before and transfer it to the same column. Collect this eluate in the same Craig tube. Evaporate the methylene chloride solution using a warm sand bath in the **hood** under a gentle stream of nitrogen gas. Recrystallize the solid amide product using the Craig tube. Dissolve the material in about 0.5 mL of ethanol, add water (dropwise) until the solution becomes cloudy, cool the Craig tube in an ice bath, and collect the crystals in the usual manner. Dry the crystalline amide on a porous clay plate and determine the melting point.

Anilides

In a 3.0-mL conical vial containing a magnetic spin vane, and equipped with an air condenser, place 5 drops of aniline and 10 drops of methylene chloride. Cool the solution in an ice bath and transfer the acid chloride solution (prepared above) via Pasteur pipet, *dropwise*, with stirring, to the aniline solution in the **hood**. *It is convenient to make this addition down the neck of the condenser.* After the addition is complete, remove the ice bath and stir the mixture for an additional 10 min.

Transfer the methylene chloride layer to a 10 × 75-mm test tube, and wash it with 0.5 mL of H_2O, 0.5 mL of 5% aqueous HCl, 0.5 mL of 5% aqueous NaOH, and finally, 0.5 mL of H_2O. For each washing, shake the test tube and remove the top aqueous layer by Pasteur filter pipet. Transfer the resulting wet methylene chloride layer to a Pasteur filter pipet containing 200 mg of anhydrous sodium sulfate. Collect the eluate in a Craig tube containing a boiling stone. Rinse the original test tube with an additional 10 drops of methylene chloride. Collect this rinse and pass it through the same column. Both eluates are combined.

Evaporate the methylene chloride solvent on a warm sand bath under a gentle stream of nitrogen gas in the **hood**. Recrystallize the crude anilide from an ethanol–water mixture using the Craig tube. Dissolve the material in about 0.5 mL of ethanol, add water (dropwise) to the cloud point, cool in an ice bath, and collect the crystals in the usual manner. Dry the purified derivative product on a porous clay plate, and determine its melting point.

Toluidides

$$R-\overset{\overset{\cdot\cdot}{\overset{\cdot\cdot}{O}}}{\underset{}{C}}-Cl + 2\,H_2\overset{\cdot\cdot}{N}-\langle\!\!\!\bigcirc\!\!\!\rangle-CH_3 \longrightarrow$$

$$R-\overset{\overset{\cdot\cdot}{\overset{\cdot\cdot}{O}}}{\underset{}{C}}-\underset{\underset{H}{|}}{\overset{\cdot\cdot}{N}}-\langle\!\!\!\bigcirc\!\!\!\rangle-CH_3 + CH_3-\langle\!\!\!\bigcirc\!\!\!\rangle-NH_3{}^+, Cl^-$$

The same procedure described for the preparation of anilides is used, except that *p*-toluidine replaces the aniline.

ALCOHOLS[15]

Phenyl- and α-naphthylurethanes (phenyl- and α-naphthylcarbamates)

$$Ar-\overset{\cdot\cdot}{N}=C=\overset{\cdot\cdot}{O}\!: + R-\overset{\cdot\cdot}{\underset{\cdot\cdot}{O}}-H \longrightarrow Ar-\underset{\underset{H}{|}}{\overset{\cdot\cdot}{N}}-\overset{\overset{\cdot\cdot}{O}:}{\underset{}{C}}-\overset{\cdot\cdot}{\underset{\cdot\cdot}{O}}-R$$

Isocyanate   Urethane

Note

For the preparation of these derivatives, the alcohols must be anhydrous. Water hydrolyzes the isocyanates to produce arylamines that react with the isocyanate reagent to produce high-melting, disubstituted ureas.

In a 3.0-mL conical vial containing a boiling stone and equipped with an air condenser protected by a calcium chloride drying tube place 15 mg of an anhydrous alcohol or phenol. Remove the air condenser from the vial and add 2 drops of phenyl isocyanate or α-naphthyl isocyanate. Replace the air condenser immediately. If the unknown is a phenol, add 1 drop of pyridine in a similar manner.

---

[15]See Table 10.3.

Caution

> This addition must be done in the **hood.** The isocyanates are lachrymators! Pyridine has the characteristic strong odor of an amine.

If a spontaneous reaction does not take place, heat the vial at about 80–90 °C, using a sand bath, for a period of 5 min. Then cool the reaction mixture in an ice bath. It may be necessary to scratch the sides of the vial to induce crystallization. Collect the solid product by vacuum filtration, using a Hirsch funnel, and purify it by recrystallization from ligroin. For this procedure, place the solid in a 10 × 75-mm test tube and dissolve it in 1.0 mL of warm (60–80 °C) ligroin. If diphenyl (or dinaphthyl) urea is present (formed by reaction of the isocyanate with water), it is insoluble in this solvent. Transfer the warm ligroin solution to a Craig tube using a Pasteur filter pipet. Cool the solution in an ice bath and collect the resulting crystals in the usual manner. After drying the product on a porous clay plate, determine the melting point.

3,5-Dinitrobenzoates

3,5-Dinitrobenzoyl
chloride

Note

> The dinitrobenzoyl chloride reagent tends to hydrolyze on storage to form the corresponding carboxylic acid. Check its melting point before use (3,5-dinitrobenzoyl chloride, mp = 74 °C; 3,5-dinitrobenzoic acid, mp = 202 °C).

In a 3.0-mL conical vial containing a boiling stone, and equipped with an air condenser protected by a calcium chloride drying tube, place 25 mg of pure 3,5-dinitrobenzoyl chloride and two drops of the unknown alcohol. Heat the mixture to about 10 °C below the boiling point of the alcohol (but not over 100 °C) on a sand bath for a period of 5 min. Cool the reaction mixture, add 0.3 mL of water, and then place the vial in an ice bath to cool. Collect the solid ester by vacuum filtration, using a Hirsch funnel, and wash the filter cake with three 0.5-mL portions of 2% aqueous sodium carbonate (Na_2CO_3) solution, followed by 0.5 mL of water. Recrystallize the solid product from an ethanol–

water mixture using a Craig tube. Dissolve the material in about 0.5 mL of ethanol, add water (dropwise) until the solution is just cloudy, cool in an ice bath, and collect the crystals in the usual manner. After drying the product on a porous clay plate, determine the melting point.

ALDEHYDES AND KETONES[16]

2,4-Dinitrophenylhydrazones

2,4-Dinitrophenylhydrazine   A 2,4-dinitrophenylhydrazone

The procedure outlined in the Classification Test Section for aldehydes and ketones (p. 275) is used. Since the derivative to be isolated is a solid, it may be convenient to run the reaction in a 3-mL vial or in a small test tube. Double the amount of the reagents used. If necessary, the derivative can be recrystallized from 95% ethanol.

The procedure is generally suitable for the preparation of phenylhydrazone and *p*-nitrophenylhydrazone derivatives of aldehydes and ketones.

Semicarbazones

Semicarbazide   A semicarbazone

In a 3.0-mL conical vial place 12 mg of semicarbazide hydrochloride, 20 mg of sodium acetate, 10 drops of water, and 12 mg of the unknown carbonyl compound. Cap the vial, shake vigorously, vent, and allow the vial to stand at room temperature until crystallization is complete (varies from a few minutes to several hours). Cool the vial in an ice bath if necessary. Collect the crystals by vacuum filtration, using a Hirsch funnel, and wash the filter cake with 0.2 mL of cold water. Dry the crystals on a porous clay plate. Determine the melting point.

---

[16]See ⟨⟩ Tables 10.4 and 10.5.

AMINES[17]

Primary and secondary amines: Acetamides

$$R-\ddot{N}H_2 + \begin{matrix} CH_3-C \overset{\ddot{O}:}{} \\ :\ddot{O}: \\ CH_3-C \underset{\ddot{O}:}{} \end{matrix} \xrightarrow{NaOAc} CH_3-\overset{:\ddot{O}:}{\underset{H}{\overset{\|}{C}}}-\ddot{N}-R + CH_3-C \overset{\ddot{O}:}{\underset{\ddot{O}H}{}}$$

In a 3.0-mL conical vial equipped with an air condenser, place 20 mg of the unknown amine, 5 drops of water, and 1 drop of concentrated hydrochloric acid.

In a small test tube, prepare a solution of 40 mg of sodium acetate trihydrate dissolved in 5 drops of water. Stopper the solution and set it aside for use in the next step.

Warm the solution of amine hydrochloride to about 50 °C on a sand bath. Then cool it, and add 40 μL of acetic anhydride in one portion (in the **hood**) through the condenser by aid of a 9-in. Pasteur pipet. In like manner, *immediately* add the sodium acetate solution (prepared previously). Swirl the contents of the vial to ensure complete mixing.

Allow the reaction mixture to stand at room temperature for about 5 min, and then place it in an ice bath for an additional 5–10 min. Collect the white crystals by vacuum filtration, using a Hirsch funnel, and wash the filter cake with two 0.1-mL portions of water. The product may be recrystallized from ethanol–water using the Craig tube, if desired. Dry the crystals on a porous clay plate and determine the melting point.

Primary and secondary amines: Benzamides

$$R-\ddot{N}H_2 + \langle\!\!\!\bigcirc\!\!\!\rangle-\overset{:\ddot{O}:}{\overset{\|}{C}}-Cl \xrightarrow{NaOH} \langle\!\!\!\bigcirc\!\!\!\rangle-\overset{:\ddot{O}:}{\underset{H}{\overset{\|}{C}}}-\ddot{N}-R + NaCl + H_2O$$

In a 3.0-mL conical vial in the **hood** place 0.4 mL of 10% aqueous NaOH solution, 25 mg of the amine, and 2–3 drops of benzoyl chloride. Cap and shake the vial over a period of about 10 min. Vent the vial periodically to release any pressure buildup.

Collect the crystalline precipitate by vacuum filtration, using a Hirsch funnel, and wash the filter cake with 0.1 mL of dilute HCl followed by 0.1 mL of water. It is generally necessary to recrystallize the material from methanol or aqueous ethanol using the Craig tube. Dry the product on a porous clay plate and determine the melting point.

---

[17]See ⚫ Tables 10.6 and 10.7.

Primary, secondary, and tertiary amines: Picrates

$R_3N: +$ [structure of Picric acid: benzene ring with :ÖH at top, O_2N and NO_2 ortho, NO_2 para] $\longrightarrow$ $R_3\overset{+}{N}H$, [structure of Picrate salt: benzene ring with :Ö:⁻ at top, O_2N and NO_2 ortho, NO_2 para]

Picric acid   Picrate salt

In a 3.0-mL conical vial containing a boiling stone and equipped with an air condenser, place 15 mg of the unknown amine and 0.3 mL of 95% ethanol.

Note

If the amine is not soluble in the ethanol, shake the mixture to obtain a saturated solution and then transfer this solution, using a Pasteur filter pipet, to another vial.

Now add 0.3 mL of a saturated solution of picric acid in 95% ethanol.

Caution

Picric acid explodes by percussion or when rapidly heated.

Heat the mixture at reflux, using a sand bath, for about 1 min and then allow it to cool slowly to room temperature. Collect the yellow crystals of the picrate by vacuum filtration, using a Hirsch funnel. Dry the material on a porous clay plate and determine the melting point.

ACID CHLORIDES AND ANHYDRIDES[18]

Amides

$$R-\overset{\cdot\overset{\cdot\cdot}{O}\cdot}{\underset{\|}{C}}-Cl + 2\,NH_3 \longrightarrow R-\overset{\cdot\overset{\cdot\cdot}{O}\cdot}{\underset{\|}{C}}-NH_2 + NH_4Cl$$

---

[18]See 🌐 Table 10.8.

In a 10×75-mm test tube, place 0.4 mL of ice cold, concentrated ammonium hydroxide solution. To this solution, in the **hood**, slowly add, with shaking, about 15 mg of the unknown acid chloride or anhydride. Stopper the test tube and allow the reaction mixture to stand at room temperature for about 5 min. Collect the crystals by vacuum filtration, using a Hirsch funnel, and wash the filter cake with 0.2 mL of ice-cold water. Recrystallize the material, using a Craig tube, from water or an ethanol–water mixture. Dry the purified crystals on a porous clay plate and determine the melting point.

AROMATIC HYDROCARBONS[19]

Picrates

Picric acid   Picrate complex

The procedure outlined on page 296 is used to prepare these derivatives.

NITRILES[20]

Hydrolysis to amides

Conversion of nitriles to water-insoluble amides, by hydrolysis with alkaline hydrogen peroxide, is a possible method of characterization for these compounds. It is especially useful for aromatic nitriles.

In a 5-mL conical vial containing a magnetic spin vane, weigh and place about 50 mg of the nitrile and 500 μL of a 1 M NaOH solution. Cool the mixture in a water bath and, with stirring, add dropwise 500 μL of 12% H_2O_2 solution. Attach the vial to an air condenser and warm the solution on a sand bath while stirring at 50–60 °C for approximately 45 min. Add 1–2 mL of cold water to the cooled reaction mixture, and then collect the solid amide

[19]See ⬤ Table 10.9.
[20]See ⬤ Table 10.13.

by vacuum filtration. Wash the product with two 1-mL portions of cold water, and recrystallize the amide from aqueous ethanol using the Craig tube. Dry the solid and determine the melting point.

PHENOLS[21]

α-Naphthylurethanes (α-Naphthylcarbamates)

The procedure outlined under Alcohols: Phenyl-, and α-Naphthylurethanes is used to prepare these derivatives (p. 292).

Bromo derivatives

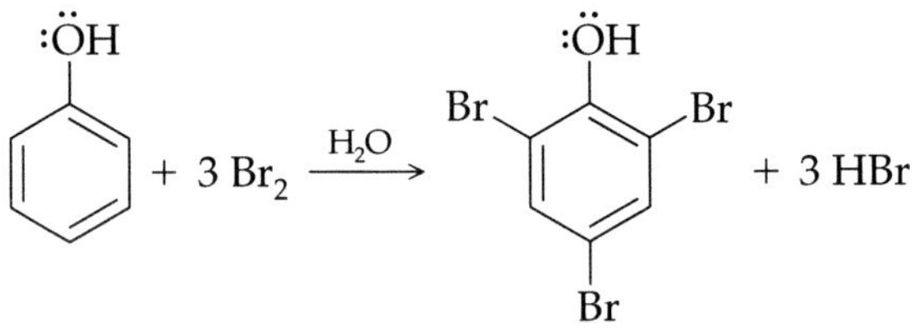

*INSTRUCTOR PREPARATION. The brominating reagent is prepared by adding 1.0 mL (3 g) of bromine in the **hood** to a solution of 4.0 g of KBr in 25 mL of water.*

In a 1.0-mL conical vial, place 10 mg of the unknown phenol followed by 2 drops of methanol and 2 drops of water. To this solution, in the **hood**, add 3 drops of brominating reagent from a Pasteur pipet.

Continue the addition (dropwise) until the reddish-brown color of bromine persists. Now add water (4 drops), cap the vial, shake, vent, and then allow it to stand at room temperature for 10 min. Collect the crystalline precipitate by vacuum filtration using a Hirsch funnel and wash the filter cake with 0.5 mL of 5% aqueous sodium bisulfite solution. Recrystallize the solid derivative from ethanol, or from an ethanol–water mixture, using a Craig tube. Dissolve the material in about 0.5 mL of ethanol, add water until it becomes cloudy, cool in an ice bath, and collect the crystals in the usual manner. Dry the purified product on a porous clay plate and determine the melting point.

ALIPHATIC HYDROCARBONS, HALOGENATED HYDROCARBONS, AMIDES, NITRO COMPOUNDS, ETHERS, AND ESTERS[22]

These compounds do not give derivatives directly, but are usually converted into another material that can then be derivatized. The procedures are, for the

---

[21]See 🌐 Table 10.10.
[22]See 🌐 Tables 10.11, 10.12, and 10.14–10.17.

most part, lengthy, and frequently give mixtures of products. It is recommended that compounds belonging to these classes be primarily identified using spectroscopic methods. Measurement of their physical properties is also of utmost importance.

QUESTIONS

10-1. The following six substances have approximately the same boiling point and are all colorless liquids. Suppose you were given six unlabeled bottles, each of which contained one of these compounds.
Explain how you would use simple chemical tests to determine which bottle contained which compound.

Ethanoic acid	Toluene
Propyl butanoate	Diisobutylamine
1-Butanol	Styrene

10-2. A colorless liquid (C_4H_6O) with a boiling point of 81 °C was found to be soluble in water and also in ether. It gave a negative test for the presence of halogens, sulfur, and nitrogen. It did, however, give a positive test with the Baeyer reagent and also gave a positive test with the 2,4-dinitrophenylhydrazine reagent. It gave negative results when treated with ceric nitrate solution and with Tollens reagent. Treatment with ozone followed by hydrolysis in the presence of zinc gave formaldehyde as one of the products.
What is the structure and name of the colorless liquid?

10-3. A colorless liquid, compound A (C_3H_6O), was soluble in water and ether, and had a boiling point of 94–96 °C. It decolorized a Br_2–CH_2Cl_2 solution and gave a positive ceric nitrate test. On catalytic hydrogenation it formed compound B (C_3H_8O), which did not decolorize the above bromine solution, but did give a positive ceric nitrate test. Treatment of compound A with ozone, followed by hydrolysis in the presence of zinc, gave formaldehyde as one of the products. Compound A formed an α-naphthylurethane with a melting point of 109 °C.
What are the names and structures of compounds A and B?

10-4. A compound of formula $C_{14}H_{12}$ gave a positive Baeyer test and burned with a yellow, sooty flame. Treatment with ozone followed by hydrolysis in the presence of zinc gave formaldehyde as one of the products. Also isolated from the ozonolysis reaction was a second compound, $C_{13}H_{10}O$, which burned with a yellow, sooty flame, and readily formed a semicarbazone with a melting point of 164 °C. The ^1H NMR spectrum of this compound ($C_{13}H_{10}O$) showed only complex multiplets that were near 7.5 ppm; the fully ^1H-decoupled ^{13}C NMR spectrum showed only 5 peaks.
What are the structures and names of the two compounds?

10-5. Compound A ($C_7H_{14}O$) burned with a yellow, nonsooty flame and did not decolorize a bromine–methylene chloride solution. It did give a positive 2,4-dinitrophenylhydrazine test, but a negative Tollens test. Treatment of the compound with lithium aluminum hydride followed by neutralization with acid, produced compound B, which gave a positive

Lucas test in about 5 min. Compound B also gave a positive ceric nitrate test. The ^{1}H NMR spectrum for compound A gave the following data:

1.02 ppm	9H, singlet
2.11 ppm	3H, singlet
2.31 ppm	2H, singlet

Give suitable structures for compounds A and B.

10-6. A friend of yours, who is a graduate student attempting to establish the structure of a chemical species from field clover, isolated an alcohol that was found to have an optical rotation of +49.5°. Chemical analysis gave a molecular formula of $C_5H_{10}O$. It was also observed that this alcohol readily decolorized Br_2–CH_2Cl_2 solution. On this basis, the alcohol was subjected to catalytic hydrogenation and it was found to absorb 1 mol equivalent of hydrogen gas. The product of the reduction gave a positive ceric nitrate test, indicating that it, too, was an alcohol. However, the reduced compound was optically inactive.

Your friend has come to you for assistance in determining the structures of the two alcohols. What do you believe are the structures?

10-7. An unknown compound burned with a yellow, nonsmoky flame and was found to be insoluble in 5% sodium hydroxide solution but soluble in concentrated sulfuric acid. Measurement of its boiling point gave a range of 130–131 °C. Combustion analysis gave a molecular formula of C_5H_8O. It was found to give a semicarbazone with a melting point of 204–206 °C. However, it gave a negative result when treated with Tollens reagent and it did not decolorize the Baeyer reagent. It also gave a negative iodoform test.

Identify the unknown compound.

10-8. An unknown organic carboxylic acid, mp = 139–141 °C, burned with a yellow, sooty flame. The sodium fusion test showed that nitrogen was present. It did not react with *p*-toluenesulfonyl chloride, but did give a positive test when treated with 5% aqueous ferrous ammonium sulfate solution, acidified with 3 M H_2SO_4, and then followed by methanolic KOH solution. A 200-mg sample of the acid neutralized 12.4 mL of 0.098 M sodium hydroxide solution.

Identify the acid.

Does your structure agree with the calculated equivalent weight?

10-9. An unknown organic liquid, compound A, was found to burn with a yellow, sooty flame and give a positive Lucas test (~5 min). Upon treatment with sodium dichromate-sulfuric acid solution it produced compound B, which also burned with a yellow, sooty flame. Compound B gave a positive 2,4-dinitrophenylhydrazine test, but a negative result when treated with the Tollens reagent. However, compound B did give a positive iodoform test.

The ^{1}H NMR spectrum for compound A showed the following:

1.4 ppm	3H (doublet)
1.9 ppm	1H (singlet)
4.8 ppm	1H (quartet)
7.2 ppm	5H (complex multiplet)

Give the structures and suitable names for compounds A and B.

10-10. A hydrocarbon, compound A (C_6H_{10}), burned with a yellow, almost nonsmoky flame. On catalytic hydrogenation over platinum catalyst it absorbed 1 mol of hydrogen to form compound B. It also decolorized a Br_2–CH_2Cl_2 solution to yield a dibromo derivative, compound C. Ozonolysis of the hydrocarbon gave only one compound, D. Compound D gave a positive iodoform test when treated with iodine–sodium hydroxide solution. On treatment of compound D with an alcoholic solution of silver ammonium hydroxide, a silver mirror was formed within a few minutes.

Identify the hydrocarbon A and compounds B–D.

10-11. A high-boiling liquid, bp = 202–204 °C, burns with a yellow, sooty flame. Sodium fusion indicates that halogens, nitrogen, and sulfur are not present. The compound is not soluble in water, dilute sodium bicarbonate solution, or dilute hydrochloric acid. However, it proved to be soluble in 5% aqueous sodium hydroxide solution. The compound gives a purple color with ferric chloride solution and a precipitate when reacted with bromine-water. Treatment with hydroxylamine reagent did not give a reaction, but a white precipitate was obtained when the compound was treated with α-naphthylisocyanate. On drying, this white, solid derivative had an mp = 127–129 °C.

Identify the original liquid and write a structure for the solid derivative. After identifying the unknown liquid, can you indicate what might be the structure of the precipitate obtained on reaction with bromine?

10-12. A colorless liquid, bp = 199–201 °C, burns with a yellow, sooty flame. The sodium fusion test proved negative for the presence of halogens, nitrogen, and sulfur. The compound was not soluble in water, 5% aqueous sodium hydroxide, or 5% hydrochloric acid. However, it dissolved in sulfuric acid with evolution of heat. It did not give a precipitate with 2,4-dinitrophenylhydrazine solution, and it did not decolorize bromine–methylene chloride solution. The unknown liquid did give a positive hydroxamate test and was found to have a saponification equivalent of 136.

Identify the unknown liquid

10-13. Your friend of Question 10-6 still needs your help. A week later a low-melting solid, compound A, was isolated, which combustion analysis showed had composition $C_9H_{10}O$. The substance gave a precipitate when treated with 2,4-dinitrophenylhydrazine solution. Furthermore, when reacted with iodoform reagent, a yellow precipitate of CHI_3 was observed. Acidification of the alkaline solution from the iodoform test produced a solid material, compound B.

Reduction of compound A with $LiAlH_4$ gave compound C ($C_9H_{12}O$). Compound C also gave compound B when treated with iodoform reagent. Vigorous oxidation of compound A, B, or C with sodium dichromate–sulfuric acid solution gave an acid having an mp = 121–122 °C.

Your friend needs your assistance in determining the structures for compounds A, B, and C. Can you identify the three compounds?

10-14. An organic compound ($C_9H_{10}O$) showed strong absorption in the IR spectrum at 1735 cm^{-1} and gave a semicarbazone having a melting point of 198 °C. It burned with a yellow, sooty flame and also gave a positive iodoform test. The ^1H NMR spectrum of the compound provided the following information:

2.11 ppm	3H (singlet)
3.65 ppm	2H (singlet)
7.20 ppm	5H (complex multiplet)

Identify the unknown organic compound.

10-15. An unknown compound (A) was soluble in ether but only slightly soluble in water. It burned with a clear blue flame and combustion analysis showed it to have the molecular formula of $C_5H_{12}O$. It gave a positive test with the Jones reagent producing a new compound (B) with a formula of $C_5H_{10}O$. Compound B gave a positive iodoform test and formed a semicarbazone. Compound A on treatment with sulfuric acid produced a hydrocarbon (C) of formula C_5H_{10}. Hydrocarbon C readily decolorized a $Br_2–CH_2Cl_2$ solution, and on ozonolysis, produced acetone as one of the products.

Identify the structure of each of the lettered compounds.

10-16. Compound A (C_7H_{14}) decolorized a $Br_2–CH_2Cl_2$ solution. It reacted with $BH_3·THF$ reagent, followed by alkaline peroxide solution, to produce compound B. Compound B, on treatment with chromic acid–sulfuric acid solution, gave carboxylic acid C, which could be separated into two enantiomers. Compound A, on treatment with ozone, followed by addition of hydrogen peroxide, produced compound D. Compound D was identical to the material isolated from the oxidation of 3-hexanol with chromic acid–sulfuric acid reagent.

Identify the structures of compounds A, B, C, and D.

10-17. Compound A (C_8H_{16}) decolorized a bromine–methylene chloride solution. Ozonolysis produced two compounds, B and C, which could be separated easily by gas chromatography. Both B and C gave a positive 2,4-dinitrophenylhydrazine test. Carbon–hydrogen analysis and molecular weight determination of B gave a molecular formula of $C_5H_{10}O$. The ^1H NMR spectrum revealed the following information for B:

0.92 ppm	3H, triplet
1.6 ppm	2H, pentet
2.17 ppm	3H, singlet
2.45 ppm	2H, triplet

Compound C was a low-boiling liquid (bp 56 °C). The ^1H NMR of this material showed only one singlet.

Identify compounds A, B, and C.

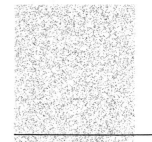

Glossary

Absorb   To take up matter (to dissolve), or to take up radiant energy.

Activated complex   An unstable combination of reacting molecules that is intermediate between reactants and products.

Activation energy   The minimum energy, $\Delta G^{\ddagger}$, necessary to form an activated complex in a reaction. Or the difference in energy levels between the ground state and transition state.

Active methylene   A methylene group with hydrogen atoms rendered acidic due to the presence of an adjacent (α) electron withdrawing group, such as a carbonyl group.

Activity (of alumina)   A measure of the degree to which alumina adsorbs polar molecules. The activity (adsorbtivity) of alumina may be reduced by the addition of small amounts of water. Thus, the amount of water present in a sample of alumina determines the activity grade. Alumina of a specific activity can be prepared by dehydrating alumina at 360 °C for about 5–6 h and then allowing the dehydrated alumina to absorb a suitable amount of water. The Brockmann scale of alumina activity is based on the amount of water (weight percent) that the alumina contains: grade I = 0%, grade II = 3%, grade III = 6%, grade IV = 10%, and grade V = 15%. For further information, see Brockmann, H.; Schodder, H. *Chem Ber.* **1941**, *74*, 73.

Adsorb   The process by which molecules or atoms (either gas or liquid) adhere to the surface of a solid.

Aliphatic   Term used to refer to nonaromatic species, such as alkanes, alkenes, alkynes.

Aliquot   A portion.

Alkaloid   A naturally occurring compound that contains a basic amine functional group. They are found particularly in plants.

Anilide   A compound that contains a C_6H_5NHCO group. An amide formed by acylation of aniline (aminobenzene).

Bimolecular reaction   The collision and combination of two reactants to give an activated complex in a reaction.

Capillary action   The action by which the surface of a liquid, where it contacts a solid, is elevated or depressed because of the relative attractions of the molecules of the liquid for each other and for the solid. It is particularly observable in capillary tubes, where it determines the ascent (descent) of the liquid above (below) the level of the liquid in which the capillary tube is immersed.

Catalyst   A substance that changes the speed of a chemical reaction without affecting the yield or undergoing permanent chemical change itself.

Characterize   To conclusively identify a compound by the measurement of its physical, spectroscopic, and other properties.

Condensation reaction   A condensation reaction is an addition reaction that produces water (or another small neutral molecule such as CH_3OH or NH_3) as a byproduct.

Dehydrohalogenation   A reaction that involves loss of HX from a halide by treatment with strong base.

Deliquescent   Liquefying by the absorption of water from the surrounding atmosphere.

Dihedral angle   The angle between two intersecting planes. In organic chemistry the term dihedral angle (or torsional angle) is used to describe the angle between two atoms (or groups) bonded to two adjacent atoms, such as H—C—C—H, and can be determined from a molecular model by looking down the axis of the bond between the two central atoms.

Dipole   The separation of charge in a bond or in a molecule with a positively and negatively charged end.

Eluant   A mobile phase in chromatography.

Eluate   The solution that is eluted from a chromatographic system.

Elute   To cause elution.

Elution   The flow, in chromatography, of the mobile phase through the stationary phase.

Emulsion   A suspension composed of immiscible drops of one liquid in another liquid (e.g., oil and vinegar in salad dressing).

Enol   A functional group composed of a hydroxyl group bonded to an alkene.

Enolate   The conjugate base of a enol, that is, a negatively charged oxygen atom bonded to an alkene. An enolate results from deprotonation α to a carbonyl group.

Enthalpy change (ΔH)   The heat lost or absorbed by a system under constant pressure during a reaction.

Entropy (S)   The randomness, or amount of disorder of a system.

Entropy change (ΔS)   The change in the amount of disorder.

Filter cake   The material that is separated from a liquid, and remains on the filter paper, after a filtration.

Free energy change (ΔG)   A predictor of the spontaneity of a chemical reaction at constant temperature. $\Delta G = \Delta H - T\,\Delta S$

Glacial acetic acid   Pure acetic acid containing less than 1% water.

Heterocycle   A cyclic molecule whose ring contains more than one kind of atom.

Heterolysis   Cleavage of a covalent bond in a manner such that both the bond's electrons end up on one of the formerly bonded atoms.

Homogeneous   Consisting of a single phase.

Homolysis   Cleavage of a covalent bond in a manner such that the bond's electrons are evenly distributed to the formerly bonded atoms.

Hydroboration   Addition of borane (BH_3) or an alkyl borane to a multiple bond.

Hydrogenation   Addition of hydrogen to a multiple bond.

Hygroscopic   Absorbs moisture.

In situ   In chemistry, the term usually refers to a reagent or other material generated directly in a reaction vessel and not isolated.

Kinetics   Referring to the rate of a reaction.

Lachrymator   A material that causes the flow of tears.

Ligroin   A solvent composed of a mixture of alkanes.

Mechanism   A complete description of how a reaction occurs.

Metabolism   The chemical processes performed by a living cellular organism.

Metabolites   The compounds consumed and produced by metabolism.

Methine   A CH group (with no other hydrogen atoms attached to the carbon atom).

Methylene   A CH_2 group (with no other hydrogen atoms attached to the carbon atom).

Mother liquor   The residual, and often impure, solution remaining from a crystallization.

Olefin   An older term for an alkene.

Optical isomers   Enantiomers. Isomers that have a mirror-image relationship.

Order of reaction   With respect to one of the reactants, the order of a reaction is equal to the power to which the concentration of that reactant is raised in the rate equation.

Oxonium ion   A trivalent oxygen cation with a full octet of electrons (e.g., H_3O^+).

Paraffins   An older name for alkanes.

Phase transfer catalysts   Agents that cause the transfer of ionic reagents between phases, thus catalyzing reactions.

Plasticizer   A substance added to a polymer to make it more flexible or to prevent embrittlement.

Polymer   A compound of high molecular mass that is built up of a large number of repeating simple molecules, or monomers.

Racemic   Consisting of an equimolar mixture of two enantiomers.

Rate equations   Equations giving the relationship between reaction rate and the concentrations of the reactants.

Reaction mechanism   The stepwise sequence of elementary reactions in an overall reaction.

Reagent   A chemical or solution used in the laboratory to detect, measure, react with, or otherwise examine other chemicals, solutions, or substances.

Reflux   The process by which all vapor evaporated or boiled from a vessel is condensed and returned to that vessel.

Rotamers   Conformational isomers that can be interconverted by rotation about one or more single bonds (e.g., *gauche* and *anti* butane).

Spontaneous process   A physical or chemical change that occurs without the net addition of energy. $\Delta G < 0$ for a spontaneous process.

Sublimation   The passing of a solid directly into vapor state without first melting.

Tare   A tared container is one whose weight has been measured. The term may also refer to the process of zeroing a balance after a container has been placed on the weighing platform.

Thermodynamics The chemical science that deals with the energy transfers and transformations that accompany chemical and physical changes.

Transition state A combination of reacting molecules that is intermediate between reactants and products.

Triturate To grind to a fine powder. (Or, washing solid organic products in a solvent in which the desired product has little solubility.)

Vapor pressure The pressure exerted by a vapor in equilibrium with a liquid or solid at a given temperature.

Ylide A neutral dipolar molecule in which negative and positive charges are on adjacent atoms.

Zwitterion A neutral molecule containing separated opposite formal charges.

Index

THE USE OF THE CORRELATION CHART OF INFRARED GROUP FREQUENCIES

This chart is by no means complete. It is scarcely possible to crowd onto one piece of paper of reasonable size all that might possibly be desired; further, as infrared continues to spread in application, new group frequencies are still being discovered. The user of such charts is urged to place his own notations for newfound correlations on the chart.

The chart has been organized into chemical group types, whose designations appear along the left-hand edge. Across the top of the chart have been indicated the various classes of molecular motions that form usable group frequencies.

The short heavy horizontal line under each group symbol indicates the extremes of the frequency region in which such groups are known to have a characteristic absorption. (Often in the past, such regions have had to be extended when a particular group is placed in molecules with less familiar groups in the vicinity, with the result that its characteristic frequencies have strayed.)

The thickness of the line is a very rough index of the intensity of this absorption. A line of tapering thickness indicates the intensity for this group to be quite variable. Often these intensity variations can be correlated with structure, but no clever way to represent this on the limited space of a chart was at hand.

An open, cross-hatched line represents a region in which there is usually more than one absorption characteristic of the particular group. For example, halogenated aliphatic hydrocarbons often have several strong, sharp absorptions in the region 950–1300 cm^{-1} to which it is difficult to ascribe specific vibrational motions, but nonetheless are characteristic of that class of molecules.

The chemical symbols are those of standard organic nomenclature. A few, perhaps, should be amplified: X = halogen *except* fluorine; M = metal; N^+—H = hydrogen attached to a positively charged nitrogen atom, as in amine salts of acids; (Σ) = a "summation" band—i.e., combination or overtone—*not* a fundamental; ϕ = phenyl ring; CJ = conjugated.

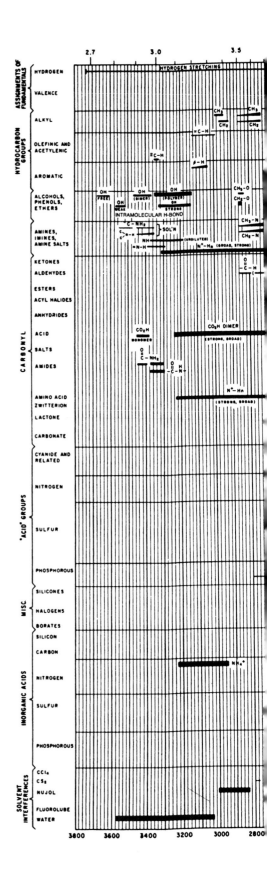

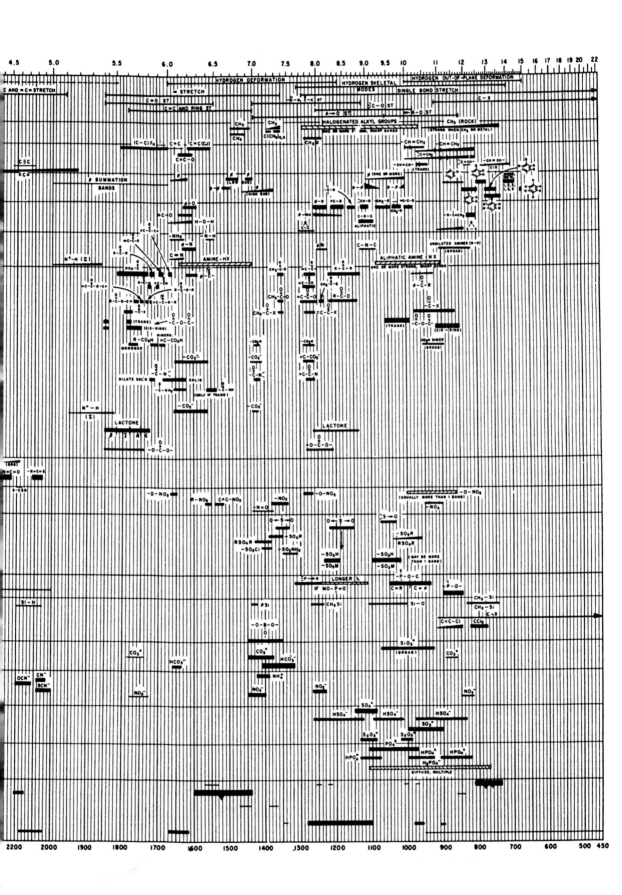

Printed in the United States
153058LV00004B/5/A

9 780471 249092